THE RE-USE OF
CONTAMINATED LAND

THE RE-USE OF CONTAMINATED LAND

A Handbook of
Risk Assessment

TOM CAIRNEY
Specialist Consultant, W. A. Fairhurst & Partners

JOHN WILEY & SONS
Chichester · New York · Brisbane · Toronto · Singapore

Copyright©1995 by Dr T. Cairney

Published in 1995 by John Wiley & Sons Ltd,
                Baffins Lane, Chichester,
                West Sussex PO19 1UD, England
                Telephone National   (01243) 779777
                                International (+44) (1243) 779777

Other Wiley Editorial Offices

John Wiley & Sons, Inc., 605 Third Avenue,
New York, NY 10158-0012, USA

Jacaranda Wiley Ltd, 33 Park Road, Milton,
Queensland 4064, Australia

John Wiley & Sons (Canada) Ltd, 22 Worcester Road,
Rexdale, Ontario M9W 1L1, Canada

John Wiley & Sons (SEA) Pte Ltd, 37 Jalan Pemimpin #05-04,
Block B, Union Industrial Building, Singapore 2057

British Library Cataloguing in Publication Data

A catalogue record for this book is available from the British Library

ISBN 0-471-94893-4

Typeset in 10/12pt Times by Dobbie Typesetting Limited, Tavistock, Devon
Printed and bound in Great Britain by Bookcraft (Bath) Ltd

This book is printed on acid-free paper responsibly manufactured from sustainable forestation,
for which at least two trees are planted for each one used for paper production.

Contents

Preface

This book has been written specifically to assist developers who reclaim contaminated land for productive re-use, and for the regulators charged with ensuring that this recycling does not expose the public or the wider environment to unacceptable risks.

Until 20 years ago there was no call for such assistance. Derelict industrial sites, with various degrees of soil contamination, were routinely recycled. Concerns over contaminated land were not recognised, and the dominant priority was to obtain development land in areas where access to "greenfield" sites was either limited or impossible. That this is the case is quite apparent when the past histories of sites in the older European industrial towns are researched; usually a very diverse sequence of activities can be identified over a century or more of uses, and periods when land had been allowed to lie vacant were both brief and rare. The legacies of land contamination from earlier occupancies seem to have had little or no disincentive, even when a particularly sensitive site use (such as domestic housing) followed directly after an exceptionally contaminating activity (such as tar refining).

Today, of course, such casual recycling of contaminated land could not be contemplated. Greater concerns for health and safety and the need to reduce environmental impairment force the risks, which could arise, to be treated seriously and be adequately reduced by appropriate site reclamation.

Currently, only in the UK does such recycling of derelict and contaminated land form a major development activity. This is because of land scarcities in a small and overpopulated island, and because governmental priorities are to bring derelict land back into productive use. In other countries, a greater caution over recycling derelict industrial land for sensitive re-uses has been apparent, though this (because of land scarcities) seems likely to decline, particularly when it can be demonstrated that no unacceptable risks will result.

Inevitably this book's risk assessment approach for acceptably safe recycling of contaminated land draws on UK experiences and regulations. Whilst this does introduce a parochial flavour, the basic principles are, hopefully, universal and applicable when recycling of contaminated land becomes a more important element of development in other countries.

The support and critical evaluations by colleagues (particularly David Hobson) are gratefully acknowledged.

T. Cairney
January 1995

PART I

NEED FOR AND BASICS OF RISK ASSESSMENTS

1

The Need for Risk Assessments when Contaminated Land is Re-Used

1.1 ORIGINS AND TRADITIONAL PERCEPTIONS OF CONTAMINATED LAND

Contaminated land is invariably typified by higher than normal contents of chemical compounds. These—at least in some circumstances—potentially can present hazards to human health, the integrity of building materials, animal and plant populations and the wider (particularly the aqueous) environment.

These enhanced chemical contents in contaminated soils originated either from earlier industrial occupancies, or were caused by the tipping of domestic or manufacturing wastes. Thus land contamination invariably is a consequence of poor controls over leakages, spillages and waste disposal.

It never is possible to list one unique suite of chemical compounds, and then claim that this always is indicative of contaminated land. This obviously is true; the range of manufacturing processes (and the consequential diversity in the fluids and solids which could have leaked or been tipped on to land) is such that distinct and different contaminant suites usually originate from different industrial activities (see Appendix I, Tables AI/1, AI/3, AI/5 and AI/6).

Because of this causal relationship between particular past activities and the types of soil contamination created, investigators can often identify discrete contaminant zones (see Chapter 3, Section 3.2.3) on larger sites, where space had allowed a number of diverse production units to co-exist.

By the standards used today to judge whether soils are "clean" or "contaminated", the contamination levels on older industrial land can often be extremely high. For example, sulphates and metals such as lead and copper can sometimes be found at concentrations of between 1 and 10% in the wastes from former glassworks and metal refineries. This can suggest that our ancestors operated especially polluting industries and showed a near-criminal casualness to the disposal of waste materials. The image that these very high contamination levels in soils can conjure up is, perhaps, that of a moonscape, with bubbling pools of oils and effluents set amongst loose piles of leachable contaminated wastes.

Whilst a few former industrial centres could have been so disorganised and polluting, in most cases the above image is certainly a gross distortion. Our predecessors did lack modern information and access to adequately sensitive chemical analytical services, but they were no less sensible than are modern industrialists. Choosing to waste or spill valuable process materials, or to complicate working areas by a poorly organised disposal of wastes would have been no recipe for commercial success.

Certainly they made use of pipes and tanks which leaked, but then so do we. Every recent demolition of modern process plants which the author has witnessed, has shown leakages and spillages, which often have caused large areas of soils to become heavily contaminated. As sub-surface investigations of old industrial sites seem not to show any higher frequencies of primary soil contamination (from leaks and spills), some other explanation for high contamination concentrations is needed.

This indicates that the real differences between the levels of soil contamination caused in the past and the distinctly lower total soil pollutions we create today, arise largely from the remedial actions which took place when an industry ceased operations and its site was to be developed for some other use. Today—with an understanding of the costs and problems of contaminated land—we will be careful to decontaminate buildings before they are demolished, and always will try to locate and empty all pipes, tanks and vats (CIRIA, 1995).

Our ancestors, however, chose the simpler approach of merely laying hardcore covers over contaminated areas, and founded their new structures on these platforms. This, in a time of more difficult and costly transportation, usually implied the use of conveniently available local materials, and these often proved to be solid wastes.

Thus slags, cinders, casting sands and such chemical wastes as gas works spent oxides were often brought in to act as cover materials. Inevitably this added a secondary, and often very significant, extra source of soil contamination.

"Remediation" of this type seems to have continued from the late 18th century, in some areas, until the mid-1970s without any public or governmental concern. This might seem surprising, since all the Western industrial nations had, by the early 20th century, devised essentially similar controls to mitigate the worst effects of industrial pollution of air and water resources, and a widespread concern for social justice had made it politically unacceptable to expose workers and local inhabitants to industrial emissions (Trevelyn, 1964). However, this concern, and the controls which stemmed from it, was restricted to the pollution of air and waters, where even untrained observers found it easy to establish cause and effect relationships, and identify the guilty agency. If— as did happen—a river (the Skerne in County Durham, UK) had been famous for its salmon yields for centuries, and then abruptly became a lifeless ferruginous drain in the mid-19th century, it was simple for local inhabitants to be sure that the loss of water life had to be due to two new collieries, pumping hundreds of thousands of gallons of saline and iron-rich mine waters into the river.

However, the environments of soils are much too complex (Failey and Bell, 1991) for such simple consequences ever to be apparent. Thus sites were used and re-used over a century or more for a range of different purposes, and a surfacing of contaminated fills (often 3 m or more in thickness) gradually built up. The developers of the period seemed not to have found this of any concern, and indeed were willing to build food-processing plants on land which previously had housed tar refinery, tannery and local authority waste tip activities.

The important perceptions were that land, in industrial centres, was scarce and that access to transportation routes, local markets and raw materials were practically important. The thought that spilled oils or solvents (for example) could pose risks to later users of a piece of land seems never to have surfaced.

1.2 RECOGNITION OF CONTAMINATED LAND

The realisation that higher than normal chemical concentrations in soils possibly could be hazardous occurred quite suddenly in the mid-1970s. The Netherlands, the UK and the USA made this discovery at much the same time, and without any prior warnings.

With hindsight, it is now apparent that it was a coincidence of two separate factors—the wider availability of sensitive chemical analytical methods, and increasing public concerns over any possible damage to the environment—which permitted the problem's recognition.

Since there was, at that time, very little real understanding of the variable fates of chemical compounds in soil environments, it was, of course, very far from clear whether or not actual hazards would arise to people or to other targets. Thus governments and their scientific advisors were faced with real difficulties. Was it necessary to react to contaminated land's recognition? (Some governments to this day have managed to avoid doing so.) What type of reaction would be reasonable and measured? Were the time scales (for any soil contaminant ill-effects to occur) so slow that crisis responses were unnecessary, or was the reverse the case?

Typical of the crucial unanswerable questions, which were current in the mid-1970s, were:

- Are there some chemical concentrations which are "acceptable" in soils? If so, how can these be confirmed quickly? If comparative health and soil quality studies have to be undertaken, then answers will not be available for years.
- Do higher concentrations of the same chemical compounds pose much more certain hazards? Can the existence of such hazards be demonstrated by surveys of sites which have been reclaimed and re-used? If so, where are the reclaimed sites which ought to be examined?

- If actual hazards seem likely to exist at some (as yet unspecified) chemical concentrations, can these be identified by relatively cheap measurements of total concentrations? Or will it be necessary to take the slower and more expensive approach of noting different compounds or isomers, which could have quite different solubilities and mobilities?
- Do easily available routes and pathways exist by which high contaminant levels in soils could impinge on human health or other environmental targets? Or is it possible that contaminants in soil environments are held so tightly that they are unlikely to pose hazards?
- Can guidance be devised for local authorities and developers, when data sources (on contaminant risks at parts per million or parts per billion levels) are so incomplete?
- Is it necessary to restrict the allowable uses to which older industrial land can be put? Or can adequately effective and affordable remediation measures (if indeed there are any which meet these criteria) solve predictable problems?
- Will it ever be possible to identify unique answers, given the range of different soil contaminants and the site-specific differences which are known to occur?

Essentially the situation was that it was technically easy to prove that soils on a site contained (perhaps) 3000 mg/kg of one contaminant and three times as high a content of another, but no one really knew the significance which ought to be given to such measurements.

Given this level of scientific ignorance, it would have been expected that the subject would have been debated within a small specialist audience, that research and advisory services would have been established to develop a fuller factual understanding, and that initial guidance and standards would have been devised and then modified as experience built up.

Any other approach—such as enacting soil quality legislation or controls—would have seemed premature and indefensible. Governments, quite properly, avoid new drains on scarce national resources, unless a real case for expenditure can be made.

In fact, the above was the pattern which developed in the UK (Harris, 1987), where a government committee (the Interdepartmental Committee on the Reclamation of Contaminated Land—ICRCL) was set up. This funded research, advised local authorities, published a range of guidance documents, and was ultimately able (1983) to issue tentative guidance for a limited number of soil contaminants. No new legislation or soil quality controls proved to be necessary, and the re-use of contaminated land remained regulated by the normal planning legislation. Since dramatic and newsworthy events did not occur, the term "contaminated land" had probably never been heard by any member of the general public until the late 1980s.

In sharp contrast, the initial Dutch experience was one of a highly publicised and extremely expensive failure. This marked public and government attitudes for years to come.

The Netherlands, in the 1960s, faced two pressing social problems—the lack of housing for a booming population and disposing of increasing quantities of domestic and industrial refuse. A "solution" to both of these difficulties, in the Lekkerkerk district, was to buy cheap farmland, fill the drainage ditches with wastes, and build homes on the platforms thus created (Hoomans and Stellingwerf, 1982). However, within a few years, abnormally frequent failures of domestic water supply pipes occurred, and investigations then revealed heavily contaminated near-surface groundwaters and volatile and mobile contaminants (including industrial solvents) below house floors.

Given the scale of the problem, and the obvious concerns of local inhabitants (some of whom did suffer from persistent poor health), immediate government action had to take place. This, in the final analysis, cost an amazingly expensive sum ($65 million) and proved to be a profoundly negative experience. Dutch officials later had no difficulty in claiming (Rorsch, 1986) that—

> soil contamination is a gigantic problem with serious consequences for man, plants (including food crops) and the abiotic environment.

—*not* because the authorities in the Netherlands had been able to achieve any fuller technical understanding than that attained in the UK, *but* because public perceptions had been so influenced by the Lekkerkerk disaster. Nationally, the mood was simply that land contamination could not be accepted in a socially responsible nation.

It is often alleged by some more extreme UK pressure groups that the Dutch, and to a lesser extent the citizens of the USA, chose to take the higher moral ground and divert vast national resources (the anticipated Netherlands expenditure on land contamination clean-ups is now said to be $27 *billion*; de Bruijn, 1992) to deal "properly" with the legacies of past industrial activities. In contrast, the UK, in this argument, is portrayed as the deliberately "dirty man of Europe", able and willing to avoid essential soil clean-up standards and costs, to protect private and institutional profits.

Like all such emotive arguments, this owes more to personal bias than to any demonstrable fact. Had the UK experienced a Lekkerkerk, or a Love Canal (Raloff, 1980) incident, it is quite certain that government would have been compelled to more proactive responses. One of the few certainties in life is that politicians need to be re-elected, and votes are never gained by ignoring public concerns.

Contaminated land is essentially identical in all the industrialised nations. There really is no difference between a disused railway siding in the Netherlands and a similar site in the UK. The same metallic, phenolic, sulphate and oily contamination will be obvious in the soils from both sites. However, public perceptions and the legislations and controls which govern land clean-up requirements and future land use *do* differ. Thus the British site can be reclaimed for less than £30/square metre and resold profitably for whatever purpose is required. The Dutch railway sidings, however, will be at least three times more expensive to reclaim, and the land's final use will be restricted. These differences arise not from better or worse reactions to contaminated soil, but to the accidents of how an ill-understood phenomenon was first perceived in the mid-1970s.

1.3 CURRENTLY AVAILABLE INFORMATION ON RISKS FROM CONTAMINATED LAND

As contaminated land became identified in other nations, these usually found it necessary to establish control strategies to protect human health (and the environment in general) when it was intended to re-use such land. As shown by Table 1.1, most countries adopted official positions intermediate between those pioneered by the Dutch and the British. Few have yet found it necessary to enact specific soil quality legislation or devise statutory soil quality standards. Instead advisory and more flexible soil quality guidelines have generally been preferred, since these allow for the site-specific differences which invariably typify contaminated land, in its often very different environmental settings. The Dutch view of multi-functionality (which requires contaminated soil to be cleaned to a standard which will permit any future land use to occur) also has been rejected elsewhere, largely because the implied costs are so enormous, and because it is very difficult to justify cleaning up the soils under (say) a car park to the pristine state suitable for agricultural purposes.

The current situation is that a large number (Table 1.2) of soil quality guidelines and standards now exist. All of these give particular chemical concentrations which are probably reasonably indicative of uncontaminated ground. A few (particularly the British ICRCL guidelines—Table 1.3) then list only a second set of distinctively higher concentrations, which are seen as showing that soils are indeed contaminated, and that remedial action is required. Others (such as the original Netherlands' standards (Table 1.4) and the similar guidelines adopted in Quebec Province and in British Columbia) employ three numerical contamination levels—the lowest values being typical of clean soils and the highest numbers indicating contamination which has to be cleaned up. The intermediate numerical values (which do not exist in the UK guidelines) are taken as the markers, which (if exceeded) show a need for further and more detailed investigation of the land. Recently (Ministry of Housing, Spatial Planning and the Environment, 1994) Netherlands policy has moved and the "A"–"B"–"C" levels have been replaced with intervention values (analogous to the former "C" levels) and much lower target values, whose attainment is the long-term national aim.

Only France, so far, has adopted the more complicated use of four numerical criteria per contaminant. These represent thresholds—the lowest values are the thresholds of "anomaly" above which contamination is possible; the increasingly higher values are the thresholds for "investigation", "treatment" and "urgency".

Table 1.1. Different national responses to land contamination

Nation/supernational grouping	Response developed
European Union	No specific soil quality directives issued. General objectives (4th Environmental Action Programme) are to identify, recover and re-use contaminated land. Protection of groundwater quality is an emerging emphasis. Directives on waste management first appeared in 1975, and have been extended in scope.
Denmark	No specific soil quality legislation enacted. Use made of 1983 Chemical Waste Sites Act for treating more hazardous sites. 1990 revision of this law extended its scope. Multi-functionality not adopted.
Germany	No specific soil quality legislation enacted. Soil Protection Act planned. States have adopted different approaches to ranking more hazardous sites. Soil quality standards not devised. Multi-functionality not adopted.
Netherlands	1983 Interim Soil Clean-Up Act introduced following the Lekkerkerk incident. Soil quality standards introduced in 1983. Multi-functionality for soil remediation standards remains official policy. Criticisms of this policy are now appearing. Soil Protection Act planned.
United Kingdom	No soil quality legislation enacted. Multi-functionality rejected as a clean-up aim. Controls exerted via planning legislation and associated environmental protection laws. Advisory soil quality guidelines exist.
Australia/New Zealand	Recognition of the phenomenon very recent. No soil quality legislation enacted. Soil quality standards not devised. Multi-functionality not adopted.
Canada	No specific soil quality legislation on a national basis. Some states (e.g. Quebec) have enacted provincial laws, similar to the Dutch pattern. National Contaminated Sites Remediation Programme enacted to establish "the polluter pays" philosophy. Interim soil quality standards in force. Provincial variations are quite significant. Regulatory control currently most severe in British Columbia.
United States of America	1976 Resource Conservation and Recovery Act in force. 1980 Comprehensive Environmental Response, Compensation and Liability Act allows federal clean-ups of contaminated sites. Progress marred by excessive legal costs and delays. Soil quality standards devised.
Other industrial nations	No official policies or phenomenon recognition as yet published.

Table 1.2. States and organisations which have devised soil quality guidelines or standards

Australia (Australia/New Zealand Environment Council)
Australia (National Health and Medical Research Council)
Australia (New South Wales State)
Australia (Victoria State)
Canadian Council of Ministers of the Environment
Canada (Alberta Environment)
Canada (British Columbia Ministry of the Environment)
Canada (Ontario Ministry of the Environment)
Canada (Quebec Ministry of the Environment)
France
Germany
Netherlands
United Kingdom
United States (Environmental Protection Agency)
United States (California Department of Public Health)
United States (Massachusetts Department of Environmental Protection)
United States (Minnesota)
United States (New Jersey Department of Environmental Protection)
United States (New York Department of Environmental Conservation)
United States (Wisconsin)
United States Army
Texaco/Shell (USA)

Irrespective of the arguable merits of these various different numerical systems (and distinct differences *do* occur in the different values set for a single contaminant, depending on the emphases adopted and the targets which the originators saw as of greatest importance, e.g. Bell *et al.*, 1989) it is a fact that seemingly comprehensive contaminant lists now exist (Angus Environmental Ltd, 1991). The existence of these could seem to take away any concerns that future risks could arise—since safety presumably would always be ensured by treating all contaminated soils to the standards set for "clean" land.

However, this would be an incorrect usage of the available soil quality standards and guidelines. It has to be remembered that none of these was devised to give assurance against all feasible risks. Most focus on potential health problems from solid contaminants found in soils and ignore leachability and mobility effects which could adversely affect the qualities of controlled water resources. Few are adequately framed to ensure that building materials will not be attacked by soil contaminants, and all—to a greater or lesser extent—can be criticised as seriously incomplete.

In more detail, the available soil standards and guidelines are obviously:

● *Partial.* A wide range of contaminants which could occur in a particular site are ignored. For example, little is specified on the toxic effects of explosive residues, yet redundant armament factories have increasingly become common, since the East–West armaments race ended.

Table 1.3(a). Tentative trigger concentrations for contaminants associated with former coal carbonisation sites (ICRCL Guidance Note 59/83 (second edition—ICRCL, 1987)
Conditions
1. This table is invalid if reproduced without the conditions and footnotes.
2. All values are for concentrations determined on "spot" samples based on an adequate site investigation carried out prior to development. They do not apply to the analysis of averaged, bulked or composited samples, nor to sites which have already been developed.
3. Many of these values are preliminary and will require regular updating. They should not be applied without reference to the current edition of the report *Problems Arising from the Development of Gas Works and Similar Sites.*
4. If all sample values are below the threshold concentrations then the site may be regarded as uncontaminated as far as the hazards from these contaminants are concerned, and development may proceed. Above these concentrations, remedial action may be needed, especially if the contamination is still continuing. Above the action concentrations, remedial action will be required or the form of development [will need to be] changed.

Contaminants	Proposed uses	Trigger concentrations: mg/kg air-dried soil	
		Threshold	Action
Polyaromatic	Domestic gardens, allotments, play areas	50	500
hydrocarbons	Landscaped areas, buildings, hard cover	1000	10 000
(notes 1,2)			
Phenols	Domestic gardens, allotments	5	200
	Landscaped areas, buildings, hard cover	5	1000
Free cyanide	Domestic gardens, allotments, landscaped areas	25	500
	Buildings, hard cover	100	500
Complex cyanide	Domestic gardens, allotments	250	1000
	Landscaped areas	250	5000
	Buildings, hard cover	250	NL
Thiocyanate (note 2)	All proposed uses	50	NL
Sulphate	Domestic gardens, allotments, landscaped areas	2000	10 000
	Buildings (note 3)	2000 (note 3)	50 000 (note 3)
	Hard cover	2000	NL
Sulphide	All proposed uses	250	1000
Sulphur	All proposed uses	5000	20 000
Acidity (pH less than)	Domestic gardens, allotments, landscaped areas	pH 5	pH 3
	Buildings, hard cover	NL	NL

Notes
NL: No limit set as the contaminant does not pose a particular hazard for this use.
1. Used here as a marker for coal tar, for analytical reasons. See *Problems Arising from the Redevelopment of Gasworks and Similar Sites*, Annex A1.
2. See *Problems Arising from the Redevelopment of Gasworks and Similar Sites* for details of analytical methods.
3. See also BRE digest 250: *Concrete in Sulphate-Bearing Soils and Groundwater.*

Table 1.3(b). Tentative trigger concentrations for selected inorganic contaminants (ICRCL Guidance Note 59/83 (second edition)—ICRCL, 1987)

Conditions

1. This table is invalid if reproduced without the conditions and footnotes.
2. All values are for concentrations determined on "spot" samples based on an adequate site investigation carried out prior to development. They do not apply to analysis of averaged, bulked or composited samples, nor to sites which have already been developed. All proposed values are tentative.
3. (a) The lower values in group A are similar to the limits for metal content of sewage sludge applied to agricultural land. (b) The values in group B are those above which phytotoxicity is possible.
4. If all sample values are below the threshold concentrations then the site may be regarded as uncontaminated as far as the hazards from these contaminants are concerned and development may proceed. Above these concentrations, remedial action may be needed, especially if the contamination is still continuing. Above the action concentration, remedial action will be required or the form of development [will need to be] changed.

Contaminants		Proposed uses	Trigger concentrations: mg/kg air-dried soil	
			Threshold	Action
Group A: Contaminants which may pose hazards to health	Arsenic	Domestic gardens, allotments	10	*
		Parks, playing fields, open space	40	*
	Cadmium	Domestic gardens, allotments	3	*
		Parks, playing fields, open space	15	*
	Chromium (hexavalent) (note 1)	Domestic gardens, allotments	25	*
		Parks, playing fields, open space	—	|
	Chromium (total)	Domestic gardens, allotments	600	*
		Parks, playing fields, open space	1000	*
	Lead	Domestic gardens, allotments	500	*
		Parks, playing fields, open space	2000	*
	Mercury	Domestic gardens, allotments	1	*
		Parks, playing fields, open space	20	*
	Selenium	Domestic gardens, allotments	3	*
		Parks, playing fields, open space	6	*

Group B: Contaminants which are phytotoxic but not normally hazards to health			
Boron (water soluble) (note 3)	Any uses where plants are to be grown (notes 2, 6)	3	*
Copper (notes 4, 5)	Any uses where plants are to be grown (notes 2, 6)	130	*
Nickel (notes 4, 5)	Any uses where plants are to be grown (notes 2, 6)	70	*
Zinc (notes 4, 5)	Any uses where plants are to be grown (notes 2, 6)	300	*

Notes

*Action concentrations will be specified in the next edition of ICRCL 59/83.

1. Soluble hexavalent chromium extracted by 0.1 M HCl at 37 °C; solution adjusted to pH 1.0 if alkaline substances present.
2. The soil pH value is assumed to be about 6.5 and should be maintained at this value. If the pH falls, the toxic effects and the uptake of these elements will be increased.
3. Determined by standard ADAS method (soluble in hot water).
4. Total concentration (extractable by $HNO_3/HClO_4$).
5. The phytotoxic effects of copper, nickel and zinc may be additive. The trigger values given here are those applicable to the "worst case": phytotoxic effects may occur at these concentrations in acid, sandy soils. In neutral or alkaline soils phytotoxic effects are unlikely at these concentrations.
6. Grass is more resistant to phytotoxic effects than are other plants and its growth may not be adversely affected at these concentrations.

Table 1.4. Standards adopted in the Netherlands for soil contaminants: A, reference value below which soils are probably uncontaminated; B, value above which there is need for further investigation; C, value above which a clean-up is indicated (Source: Moen et al., 1986)

	Substance	Concentration in soil: mg/kg dry weight			Concentration in groundwater: μg/l		
		A	B	C	A	B	C
Metals	Cr	100	250	800	20	50	200
	Co	20	50	300	20	50	200
	Ni	50	100	500	20	50	200
	Cu	50	100	500	20	50	200
	Zn	200	500	3000	50	200	800
	As	20	30	50	10	30	100
	Mo	10	40	200	5	20	100
	Cd	1	5	20	1	2.5	10
	Sn	20	50	300	10	30	150
	Ba	200	400	2000	50	100	500
	Hg	0.5	2	10	0.2	0.5	2
	Pb	50	150	600	20	50	200
Inorganic pollutants	NH (as N)	—	—	—	200	1000	3000
	F (total)	200	400	2000	300	1200	4000
	CN (total free)	1	10	100	5	30	100
	CN (total complex)	5	50	500	10	50	200
	S (total)	2	20	200	10	100	300
	Br (total)	20	50	300	100	500	2000
	PO (as P)	—	—	—	50	200	700
Aromatic compounds	Benzene	0.01	0.5	5	0.2	1	5
	Ethyl benzene	0.05	5	50	0.5	20	60
	Toluene	0.05	3	30	0.5	15	50
	Xylene	0.05	5	50	0.5	20	60
	Phenols	0.02	1	10	0.5	15	50
	Aromatics (total)	0.1	7	70	1	30	100

Polycyclic aromatic compounds (PCAs)	Naphthalene	0.1	5	50	0.2	7	30
	Anthracene	0.1	10	100	0.1	2	10
	Phenanthrene	0.1	10	100	0.1	2	10
	Fluoranthene	0.1	10	100	0.02	1	5
	Pyrene	0.1	10	100	0.02	1	5
	Benzo(a)pyrene	0.05	1	10	0.01	0.2	1
	Total PCAs	1	20	200	0.2	10	40
Chlorinated organic compounds	Aliphatic chlorinated compounds (individual)	0.1	5	50	1	10	50
	Aliphatic chlorinated compounds (total)	0.1	7	70	1	15	70
	Chlorobenzenes (individual)	0.05	1	10	0.02	0.5	2
	Chlorobenzenes (total)	0.05	2	20	0.02	1	5
	Chlorophenols (individual)	0.01	0.5	5	0.01	0.3	1.5
	Chlorophenols (total)	0.01	1	10	0.01	0.5	2
	Chlorinated PCA (total)	0.05	1	10	0.01	0.2	1
	PCB (total)	0.05	1	10	0.01	0.2	1
	EOCl (total)	0.1	8	80	1	15	70
Pesticides	Organic chlorinated (individual)	0.1	0.5	5	0.05	0.2	1
	Organic chlorinated (total)	0.1	1	10	0.1	0.5	2
	Pesticides (total)	0.1	2	20	0.1	1	5
Other pollutants	Tetrahydrofuran	0.1	4	40	0.5	20	60
	Pyridine	0.1	2	20	0.5	10	30
	Tetrahydrothiophene	0.1	5	50	0.5	20	60
	Cyclohexanone	0.1	6	60	0.5	15	50
	Styrene	0.1	5	50	0.5	20	60
	Fuel	20	100	800	10	40	150
	Mineral oil	100	1000	5000	20	200	600

Table 1.5. Speciation ranges occurring in metals (zinc as an example)

Chemical form	Example
Simple hydrated metal ion	$Zn(H_2O)^{6+}$
Simple inorganic metal ion	$Zn(H_2O)_5Cl$
Stable inorganic complexes	Zn_5
Simple organic complexes	Zn in composts
Absorbed on inorganic colloids	Zn in clays
Absorbed on organic colloids	Zn in humus
Free metal particles	

- *Incomplete* in the range of situations they address. Some metals (such as copper) are included, but usually only in so far as harm to plant growth could occur. Yet some copper salts, if inhaled or ingested (Barry, 1991), can be distinctly harmful to human health. Despite this, guidelines, such as those valid in the UK, give no guidance to allow for this risk.
- Ignore the effects of *metal speciation*. Metals occur in a wide range of possible combinations, in which their toxicities and bio-availabilities vary widely (Table 1.5). The sampling and analyses of materials, from a particular site, are seldom adequate to allow this type of speciation detail to be assessed. Cadmium, for example, does have serious health effects if the metal can be concentrated in the human kidneys or liver, but can occur in insoluble and non bio-available forms (in—for example—alkaline or hard water conditions) or as the extremely bio-available complexes which exist at very low or very high pHs. Since soil contamination guidelines and standards choose not to go into such detail, it is unsurprising that investigators are content merely to establish the total cadmium contents of soils, and not whether the metal is, or is not, in a more hazardous form.
- Ignore the effects of those *environmental factors* (acidity, reducing or oxidising environments, the presence of clay minerals, organic contents of soils, etc.) which can significantly govern whether or not some contaminants are more bio-available and mobile.
- Ignore the *interactions* which can occur between different contaminants. If more than one metal exists in a soil at high concentrations diverse effects are possible. Sometimes one metal's toxicity is increased because of the presence of another contaminant (synergistic effect); in other situations the reverse occurs as competition reduces the overall contaminant effect (antagonistic effect). In yet other conditions, interaction does not take place, and simple additive effects arise. As if this complexity were not enough, it is known (from studies on simple aquatic life forms) that combinations of the same two metals can exhibit quite distinct effects on different targets (Table 1.6). This depth of knowledge simply is not available when higher life forms are exposed to contaminant attack, and so cannot be included in any current soil standards and guidelines.

Table 1.6. Typical inter-reaction complexities

Metals occurring	Effects of the inter-reactions	Targets affected
Nickel and zinc	Synergism Antagonism Additive	Bacteria Fungi Cyanobacteria
Cadmium and nickel	Synergism Antagonism	Bacteria Algae
Nickel and lead	Antagonism Additive	Fungi Algae

The essential fact is that—despite the apparent comprehensiveness of some soil quality standards and guidelines—they are based on the best available, but very incomplete, scientific data. Many of the critical contaminant concentrations, cited in these soil quality listings, have necessarily been predicted from experiments or accidents in which high contaminant doses and exposure durations occurred. The risks of exposure to the minute concentrations which actually typify contaminated land, however, cannot yet be measured by such animal experiments or epidemiological studies (Griffin, 1988). Thus it has to be realised that the information contained in soil quality standards and guidelines is *not* intended to be absolute fact, but only the best estimates currently available.

In situations where precise causative links between an event and an outcome are unclear, the usual solution is to amass enough information and devise that empirical relationship which best fits the data. This, of course, is the approach adopted in devising the sort of travel/fatal accident risk assessments familiar to most readers (Table 1.7). With the risks from contaminated land, however, even this approach is denied. Simply there is almost no information that any risks actually have occurred.

Given the limitations of scientific information and the impossibility of devising adequate empirical relationships between the presence of soil contaminants and actual harm to people or other targets, a current enthusiasm is to utilise the power of modern computer systems, and model the fates of contaminants in soils. The more complex of these models allow for a range of pathways between contaminants and targets, and can include a wide body of information on contaminant mobilities, toxicities and solubilities (Clark, 1993). Thus they are undoubtedly very useful in assessing different situations and evaluating possible remediation strategies.

However, can it ever be claimed that the risk assessments derived by such complex models are actually correct?

In any relatively complex environmental situation, the necessary computational approach is always to devise a suitable set of mathematical expressions (thought to represent the processes which will occur) then load in a body of collected

Table 1.7. Typical travel/fatal accident analysis (Department of Transport, 1991)

Mode of travel	Passenger deaths per billion (10^9) kilometres travelled
Rail	0.45
Bus	1.20
Scheduled airline service	1.40
Car/taxi	7.00
Motor cycle	359.00

information, and predict the likely result. Usually this (*if* the actual result is known) will reveal errors and unacceptable inaccuracies, since the chosen computational accuracies can be inappropriate (Figure 1.1). Corrections to the model have then to be made before it can sensibly be used as a predictive tool.

However, in a situation where the actual result (such as the load of soluble contamination which could ooze into a lake along a long soil–water interface) is not known, the essential stage of calibrating a mathematical model's accuracy is simply not possible. Thus it is extremely difficult, despite the complexity and sophistication of some modern computer models of contaminated land's effects on possible targets, to give practical credence to the risk assessment predictions which are often claimed.

The existence of various soil quality guidelines and standards is obviously useful, and gives a framework which allows the remediation of contaminated land to proceed. However, this guidance is not enough to adequately typify any particular piece of contaminated land, or establish whether or not future risks will occur.

Some attempt to carry out more comprehensive risk assessments will usually be necessary, and this will have to take note, not only of the particular site-specific environment and the existence of targets around or on the contaminated site, but of the regulations and legislation which locally are in force. As Table 1.1 indicates, control regulations vary from country to country, and this obviously compels any detailed consideration of these restraints to focus on one specific body of controls and legislation, if the scale of potential future risks are to be properly established. A particular factor which forces this approach is that legislation in some countries (e.g. the USA) is highly prescriptive and leaves little room for interpretation either by control bodies or by the courts. Elsewhere (e.g. in the UK), finer interpretation of what a law really means is made by the courts, and previous case law judgements are often extremely important.

1.4 RISKS DUE TO A SPECIFIC FRAMEWORK OF REGULATIONS AND CONTROLS

When contaminated land is reclaimed for re-use, a wide range of risks is possible. These could occur,

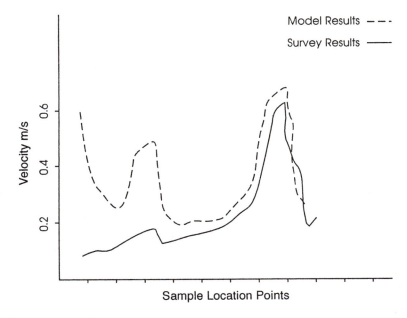

Figure 1.1. Predicted and measured water velocity (spring tide conditions)

- when unreclaimed land is first purchased, if pollution of air and water resources is occurring, and exceeds acceptable levels,
- when development approval for the reclamation method, and the planned after-use of the land, is sought,
- during reclamation, if public health, air qualities or water resources are put at risk, and
- after redevelopment, if future purchasers of the land find that conditions are not to the standards they had anticipated.

Differences in the local armoury of controls and legislations which are in force will have meaningful effects on these and other possible risks.

The UK differs from other industrial countries in having developed a large private-sector land reclamation industry. This routinely recycles several hundred formerly contaminated sites each year for profitable resale (mainly for domestic housing and light industrial uses). Given,

- the lack of statutory soil clean-up legislation or compulsory soil quality standards in the UK,
- the use of advisory soil quality guidelines, which permit a large degree of professional interpretation to allow for local situations,
- the fact that UK landowners need not clean-up contaminated land (though action can, in theory, be forced if wider scale environmental degradation is being caused by the state of the land),

- the reality that clean-up will usually only take place when a land use change is intended, and
- that developers have the duty to devise suitable land remediation methods (though these have to be accepted by the various control bodies),

it could be thought that UK contaminated land reclamations operate in a much less constrained control environment than is possible in other countries.

In fact, this is *not* the case.

The use of a diverse body of existing legislation (Table 1.8), to control the reclamation and allowable re-use of old industrial sites, means that the restrictions are complicated, and that a wide range of risks can be experienced by land developers.

Once a developer purchases a parcel of contaminated land, it is possible that he or she could also inherit liabilities. In particular there could be concerns over legal enforcements if the land posed a "statutory nuisance" (as defined in the Environmental Protection Act 1990). Neighbours, if adversely affected by fumes, smoke, dust, smells or by any accumulations of wastes which might be prejudicial to health, could complain to local Environmental Health Authorities. These, in turn, might require costly action, to remove the nuisances, or (if the landowner failed to do this) could initiate their own remedial actions, and reclaim their costs from the landowner.

Potentially more worrying would be the situation if an area of contaminated land were polluting surface or groundwater resources. The Water Resources Act 1991 (Section 85) makes it a specific offence to "cause, *or knowingly permit*, any direct discharge of poisonous, noxious or polluting matter to any controlled water". Thus any developer who intends to acquire a disused industrial site ought either to be sure that the land does not currently pollute water resources, or that the costs of rectifying the situation have been taken into account before any land purchase price was fixed.

Table 1.8. English* and Welsh* legislation which can affect the redevelopment of contaminated land

Town and Country Planning Act 1990
Town and Country Planning (Assessment of Environmental Effects) Regulations 1988
Building Regulations 1991
Environmental Protection Act 1990 (Part III on Statutory Nuisance)
Health and Safety at Work Act 1974
Clean Air Act 1993
Environmental Protection Act 1990 (emissions from prescribed processes)
Water Resources Act 1991
Environmental Protection Act 1990 (Part II on Waste)

*Legal constraints in Scotland and Northern Ireland differ in some details, though their overall impact is essentially the same.

As an aside, this particular aspect of English and Welsh law has been tested recently in the courts. This has revealed that, whilst those who drafted the law obviously intended it to be an offence to own land which was oozing mobile contaminants (say oily residues or landfill leachates) into streams or ground-waters, the courts do not fully support this view. If the noxious substances do enter the controlled water via a pipe (even if this pipe was installed and checked by a previous landowner's agents), the courts have agreed (Anon., 1994b) that the present landowner is legally at fault. However, where uncontrolled leakages, through soils, into controlled waters have taken place, a recent legal ruling is that this is *not* a discharge, since (in the legal sense) a discharge seemingly has to have a unique point of occurrence (Anon., 1994a). Thus buying land which is oozing contaminated fluids into a stream need not be as obvious a breach of the law as the wording of the Water Resources Act could suggest.

The use of case law judgements to refine and alter the legislation, which Parliament believed it had enacted, is a complexity which means that control bodies in England and Wales normally are reluctant to initiate legal actions until all other avenues of discussion and compromise have been exhausted.

Planning approval is the critical first hurdle when it is intended to recycle old industrial land. Land contamination, or the potential for this, is a "material consideration" which local authorities have to consider before allowing any particular land re-use. If contamination does exist, the authorities can reject a planning application, or restrict the allowable land uses, or impose specific development conditions. Usually the last option is the outcome, and conditions that are common are:

- the requirement for a fuller site investigation,
- the need to revise the planned reclamation method, especially if consultees such as the National Rivers Authority and the regional Waste Regulatory Authority have voiced concerns, and
- quality achievement checks on the final land remediation.

Such planning conditions, however, have to be fairly and reasonably related to the proposed development, and have to be capable of practical enforcement. Should a local authority's conditions be seen as unreasonable, a developer can always appeal to the Secretary of State for the Environment.

Even after planning consent has been granted, a further safeguard exists in the Building Control regulations. These permit inspectors to take a direct interest in "substances found on or in the ground to be covered by the building" (Building Regulations—*Approved Document C2*—1991) and pose a further risk to a developer. Since the wording of these regulations allows inspectors to show an interest in any substance which "is, or *could become* toxic, corrosive, explosive, flammable or radioactive, and so become a danger to health and safety," there can be considerable scope for building control officers to reject reclamations, which the planning authorities have already approved.

Health and Safety and Clean Air regulations will also be significant during the land reclamation period, and legal enforcement of necessary standards of public and workforce safety can occur. Should a reclamation be so poorly planned that serious health and safety concerns arise, then substantial corrective costs (including health surveillance for site operatives) could be required, and legal prosecution is likely.

In addition to these controls, strict regulations now govern the disposal and removal of waste materials, such as the contaminated fill layers, which mantle many older industrial sites.

If a developer has been careful not to acquire land which could attract the concern of the bodies charged with protecting the air and water environments; has fully satisfied the planning authorities over the adequacy of necessary land remediation; and finally has avoided exposing workpeople (or the local population, or the surrounding streams and water resources) to pollution; it could be thought that any further exposure to risks would be minimal. However, this need not be the case, as future purchasers of, or residents on or around, a reclaimed site might find that conditions are less acceptable than really are wanted, and choose to seek legal satisfaction. A range of legislation could be used in such actions:

- the Property Misdescription Act 1991, which makes false or misleading marketing claims a criminal office,
- the Misrepresentation Acts, whose breach could lead to charges of fraudulent misrepresentation,
- the Defective Premises Act 1972, which calls for buildings to be designed properly. Land contamination, below structures, could be taken to adversely affect the fitness of buildings for their designed use, and so lead to actions, and
- Common law actions.

Common law is that body of law established, not by Acts of Parliament, but by long-established principles. Whilst essentially static in nature (in that no new rights and remedies are constantly being created), it is not however stagnant. Courts do have a limited ability to extend traditional rights and remedies to accommodate new circumstances. Common law actions, over either negligence or nuisance, have traditionally been pursued by individuals at their own costs. However, recently (Anon., 1993b and c) legal aid has been granted, from national resources, to permit individuals to pursue civil claims over environmental nuisances. This trend, should it continue, will pose an especially serious risk to developers who fail to ensure that contaminated land clean-ups are to adequate standards. Obviously there are many more members of the public than there are control body inspectors, and if the public comes to believe that it is being exposed to the effects of nuisance or negligence and also has access to the funds needed to mount legal challenges, then a marked increase in litigation is predictable.

1.5 WHY RISK ASSESSMENTS ARE NECESSARY

Contaminated soils are widespread in all the longer-established industrial centres, and (at least) detract from an area's appearance. At worst, these soils could be a future source of adverse effects on human health and environmental qualities.

Thus, it makes good sense to reclaim and re-use such sites and at the same time restore the crumbling fabric of older industrial towns. The increasing scarcity of land resources in Western Europe, adds a further and practical impetus to this view.

However, when re-use of former industrial land is proposed, risks are possible. This is an inevitable consequence of the complexities which do occur. The range of alien chemical compounds which can be found in soils is extremely diverse, and the local environment around any one site will always be unique. Thus some degree of individual assessment of any particular site always has to be undertaken to identify the more feasible of the risks which could arise.

Relying on published soil quality standards (statutory requirements), or on advisory guidelines, will never give adequate safeguards from risk. These published lists are never fully comprehensive, nor indeed were they devised to encompass all possible site-specific variations.

A range of interest groups inevitably becomes involved when contaminated land is to be reclaimed:

- developers, whose interests are in minimal costs and fastest development completions,
- financial backers, who wish to maximise their profits without undue exposure to any risks,
- consultants, acting either for developers or for control bodies,
- planning authorities and control bodies (acting for local authority and environmental protection interests) whose aims are to minimise adverse impacts and avoid the public being put at risk, and
- the interested public, usually represented by local environmental pressure groups.

The perceptions and concerns of these groups obviously will differ, and conflict is near-inevitable, unless comprehensive and reliable risk assessments are offered to allow rational discussions. The reactions possible (for example) over the mention of a single contaminant's occurrence can often seem amazingly overstated to a reclamation specialist, but this often is due to no more than a less than complete assessment of the significance (which can be given to the particular contaminant's presence) being supplied.

Risks that could affect the various parties also will differ.

Developers are exposed to the greatest range of feasible risks:

- of funders refusing, or withdrawing, financial backing, because inadequate confidence has been given by the developer's proposals,

- of unknowingly purchasing land, which is so affecting the local environment, that excessive (and unbudgeted) remedial costs are necessary, or legal penalties incurred,
- of long delays in the planning application process, because the concerns of control bodies have not been adequately addressed,
- of being unable to sell on reclaimed land if potential purchasers are not able to be convinced that reclamation has been adequately completed,
- of being exposed to threats of a range of criminal or civil actions (even if these are not pursued to actual court hearings, very large defensive legal costs can be incurred), and
- of being compelled to defend criminal or civil actions in court. The penalties if guilty verdicts are handed down can be significant, and to these has to be added the loss in reputation and the adverse press publicity.

Developers' funders essentially are exposed only to two risks of significance:

- of financial loss, should a development fail to meet its financial targets, and
- of inheriting the developer's legal liabilities, in those cases where a developer is bankrupted.

A developer's consultant also only has to be concerned over a limited list of outcomes:

- of losing client confidence (and future commissions), and
- of being sued by the client, should professional negligence be thought to have been the cause of unacceptable costs, delays, or inadequacies.

In contrast, planning authorities and control bodies risk not these commercial penalties but the concerns of not adequately protecting the public or the environment (pressure groups in the UK have commenced challenging decisions of control bodies, which they believe have not fully enforced statutory law), or the fears of their decisions being challenged by a developer in the courts or on appeal to higher political authority.

Finally, the public has the most obvious of all concerns—that of being exposed to unacceptable and dangerous conditions.

These differing concerns and fears of future risk are best addressed by a process of formal and consistent risk assessment, focused on the particular conditions of an individual site.

Contaminated land is being recycled and the process inevitably will continue as long as land resources are scarce. However, if unacceptable consequences are not to occur, it is essential that the complete range of feasible risks from a particular site redevelopment are assessed. If this is done in a consistent and convincing manner, the risks of importance to the various interest groups can be quantified, and appropriate remedial actions included in the reclamation plan. Any other approach in the current state of scientific uncertainty is, quite bluntly, immoral.

2

Risk Assessment Methods

2.1 INTRODUCTION

In all contaminated land situations, risk can only be postulated if:

(a) a hazard exists,
(b) a pathway occurs via which the effects of that hazard can be transmitted, and
(c) a target or receptor is then exposed to doses of the contaminant hazard.

The causal chain of:

$$Hazard \rightarrow Pathway \rightarrow Target/Receptor$$

thus has to be unbroken. Should pathways be closed off (as occurs in well-designed encapsulations of buried contaminants) then no risks will occur, even if the contaminant present has a particularly high hazard rating.

All meaningful risk assessment methods adopt this approach.

Before considering the variations of risk assessment that are feasible, it is important to realise that *absolute certainty* will always be *illusory*, and all that can be achieved is to rank risks and ascribe a probability of occurrence for each. On occasion some assessors overstate what is attainable. The statement, for example, that "contact with this contaminant will mean that 'x' people (whether 'x' is a single number or a numerical range) will contract cancer" means either that "the particular assessor is incompetent or simply is a liar" (Rodricks, 1992).

Absolute truths are always beyond the scope of risk analysis in the complexities of land contamination.

It also is worth noting that "risk" is not a simple concept. A wide range of interest groups will inevitably exist around a contaminated land reclamation—e.g.:

● landowners
● land purchasers

- funding institutions
- mortgage providers
- insurers
- control bodies set up to protect public health and minimise environmental degradation
- individuals who live on reclaimed sites, and
- environmental pressure groups

The interests and perceptions of these different parties are unlikely to coincide, and so the importance given to a particular land contamination occurrence will seldom be consistent.

Additionally "risk" should not be viewed narrowly and only as predictable harm to health, to non-human life forms, or to the structures of houses. Financial loss and adverse publicity in many circumstances can be risks of equal concern.

Site-specific features must also be recognised as controlling all aspects of the hazard–pathway–target/receptor chain. The soil contaminants which occur, the concentrations of these substances, and whether or not the compounds present are soluble, mobile or available for uptake are all factors which vary from site to site. So the scale of possible contaminant hazard will always be site specific. Additionally, some chemical compounds are far more hazardous than are other forms of the same elements (Chapter 1, Section 1.3).

Pathways also will vary with site conditions. If only human health is taken as the risk of concern, pathways would include those leading to direct skin contact with a contaminant, breathing in air polluted with contaminated dust or with gases and vapours rising from the site, eating crops in which hazardous contaminants have become concentrated because the plants drew up contaminated soil water for their growth and development, and drinking water into which contaminated soil or water has migrated. When other risks are considered, yet other pathways have to be evaluated.

Finally the targets at risk will also differ, and will depend on the re-use to which a reclaimed site has been put, and on the populations and the wider environment which surround that site.

Thus it is inevitable that a contaminated land risk assessment always has to be a unique and separate evaluation exercise, and this obviously does impose additional (and sometimes large) costs which have to be borne by whatever party requires the assessment.

This last point is particularly important in the UK, where risk assessments are driven not by legislative requirements but by the concerns of corporations, developers and purchasers. When only private funds exist to bear assessment costs, a lesser willingness to accept high bills is usually apparent. The situation in the USA is distinctly different; there legislation and the policy of the Environmental Protection Agency (EPA) mean that risk assessment costs are a normal and acceptable consequence of cleaning up and re-using disused industrial land. So risk assessments will usually be more detailed and apparently much more comprehensive than is normally the case in Europe.

2.2 REQUIREMENTS OF ANY DEFENSIBLE CONTAMINATED LAND RISK ASSESSMENT APPROACH

Any risk assessment, which is to be convincing, has to be:

(a) *Consistent*. The same risk evaluation outcome should always be obtained, if other assessors analyse the same data gained from a particular site. Likewise (and quite a common occurrence today), if a large number of sites are being evaluated by a team of assessors (perhaps to identify where a large landowner should target scarce remediation resources (British Gas, 1992)), it has to be certain that each individual assessor will give the same weightings and significances to similar information. Currently in the UK, this requirement is satisfied less often than should be the case.

(b) *Formal*. The need for consistency in approach implies that individual assessor choices and biases must be restricted and that formal protocols and evaluation methodologies must be followed. Dramatically different evaluation emphases, however, do occur currently in the UK, and reflect the variable expertise and experience of the assessors who are active in risk assessments (Box 2.1).

(c) *Flexible*. Although particular lists of contaminants, contaminant concentrations and contaminant associations have to be used to form judgements on the potential hazards which could exist, future research could show that some contaminant hazard rankings were ill-based. This is particularly obvious in an area where the scientific data base is currently so poorly established. To avoid the need and cost of entirely repeating risk assessments, when such new information appears, it is sensible to ensure that judgemental criteria and consequences are clearly identified. If this is done, then the limiting significance of selecting a particular contamination event as especially hazardous need not be a problem, as minor revision of the risk evaluation can easily be carried out.

(d) *Comprehensive*. All feasible risk situations should be analysed. Whilst this is a self-evident requirement, it is one which many risk assessments fail to meet. In the UK too many assessors still focus only on the risks to land users and occupiers (and often do not, in fact, evaluate all such risk categories) and ignore the risks of wider environmental impairment. In a situation in which control bodies (e.g. the National Rivers Authority) have very considerable legislative and penal powers, this omission is indefensible.

(e) *Able to identify information deficiencies*. Site investigations need not invariably provide all the information which is needed to assess the risks a contaminated site could pose. Additionally, it has to be accepted that site investigations can never fully establish the factual sub-surface conditions,

Box 2.1. Abstracts from some qualitative risk assessments

Site A   Claim   "No risks from landfill gases are likely. Methane was not encountered and carbon dioxide seldom exceeded 2.1% by volume."

Actual situation   The gas concentration data on which the judgements were made were collected in a two-week period in the middle of winter. Later re-investigation (by the potential purchaser of the land) in the following summer revealed a much riskier landfill gas situation. The original assessor had patently failed to predict the usual consequence of biodegradation accelerating as soil temperature increases.

Site B   Claim   "Insurance cover will be difficult to obtain, since high concentrations of metal contamination occur at shallow depths. This will pose likely future and serious health risks to the owners of the planned houses."

Actual situation   Slags from 1850s copper refining do exist at depths of some 3.0 m. The slags were tipped into what was a building stone quarry, and now lie entirely below the near-surface groundwater table. All materials above the groundwater are uncontaminated when compared to the ICRCL guidelines. The groundwater, however, does *not* contain dissolved metals from the slags and is in fact remarkably clean. The original assessor obviously failed to consider the leaching potential of the old copper slags and so over-exaggerated the risks that exist.

Site C   Claim   "Spilled oils do occur in areas of the site's ashy fills. The underlying glacial clays occur at no deeper depth than 3.5 m and lack any pervasive groundwater. Only localised perched groundwater occurs. This glacial clay formation removes concerns that groundwater or surface water oil pollution is likely to occur."

Actual situation   Sandier lenses exist in the boulder clay and support the perched groundwaters. Whilst no site-wide groundwater, as such, exists, later water sampling and analysis revealed that high concentrations of polyaromatic hydrocarbons (PAH) ($>$ 40 mg/litre) did exist in all perched waters. This additional fact led to a hydrological investigation which indicated that PAH pollution was already entering an adjacent stream via the supposedly impermeable glacial clay. The potential land purchaser immediately withdrew his offer for the site.

particularly where decades of human activity have affected site conditions (Chapter 3). Thus it will be normal for an assessor to lack some element of information which is needed to judge risks. A practically useful risk assessment method will invariably highlight any such information deficiencies, and preferably in such a way as to persuade the parties concerned that additional minimal investigation has to be carried out. This usually will best be achieved by ascribing a worst-case risk judgement to any areas of data deficiency.

The differences in the perceptions of the parties involved with the safe re-use of reclaimed land have already been noted. Thus any risk assessments which are conducted have to be logical, convincing and extend to encompass all feasible risks.

The lack of certainty over many land contamination hazard ratings implies that formalised risk assessments are essential, since only in this way can consistency be assured and assessments revised as additional hazard information becomes available.

Finally, all of this has to be achieved at as low a cost as is possible, until such time as legislation and/or government funds permit a less cost-conscious approach.

2.3 QUALITATIVE RISK ASSESSMENTS

Common practice in the UK, currently, is to adopt a simple qualitative risk assessment approach. In this, the various risks are ranked, but are categorised only as "low", "medium" or "high".

This has the merits of relative simplicity and of recognising the current uncertainties in the hazard ratings of many contaminants. Additionally, qualitative assessments can be completed at a cost (perhaps as low as £1000 to £2000) which clients are able to accept.

Whilst some authors (e.g. Smith, 1995) accept that this approach is generally satisfactory, and in line with the site-specific nature of land contamination, it is immediately obvious that it fails to meet the criteria noted above (Section 2.2). Assessors fail to follow any consistent approach, a wide scope for personal bias is possible, and the judgemental processes, by which a particular risk is deemed to pose particularly high risks, can be quite unclear, when the site information and risk assessment report are re-analysed by a second assessor.

As a result, the usual outcome, today, is that one party's specialist consultants prepare a risk assessment, which is then disputed by other consultants engaged by another interest group. The debates, on points of disagreement between specialists, are usually extremely unproductive, since quite different (and usually undefined) judgemental processes and methods will be being applied. Quite significant errors can also be apparent (Box 2.1), and it seems obvious that current qualitative risk assessments lack the rigour and comprehensiveness which actually are necessary in practice.

2.4   QUANTIFIED RISK ASSESSMENTS

Quantified risk assessments are commonplace in North American practice (e.g. La Goy *et al.*, 1989; Marshall *et al.*, 1989), largely because the US EPA has adopted a regulatory risk assessment approach for Federally funded and similar site clean-ups.

To European readers, the level of numerical precision contained in these assessments appears extremely impressive. For example (Traves, 1992), to be informed that the lifetime risk of children and adults developing cancer would be reduced (from the levels of risk that the currently undeveloped site poses) by a factor in excess of 3000 if a derelict industrial site were to be developed for housing, offers a detailed dimension which is not available in European practice.

This USA emphasis on cancer risks, as the main risk of concern, reflects the shift in EPA interest (from minimising gross pollution of air and waters, to reducing human health risks) and mirrors public concerns over the toxicities of chemical contaminants and the national fear of cancer as a major killer. However, the decision to refuse to countenance any safe threshold for carcinogenic chemicals, and to insist on a regulatory no-threshold approach, probably makes risk assessments based on developing lifetime cancers a particularly severe test.

The USA approach is interesting, both technically and politically. Adequate governmental funds are available to allow the establishment of potency estimates for an impressively long list of chemical substances.

The potency of chloroform (for example) is defined as 0.006 units of lifetime cancer risk for each milligram of the contaminant intaken daily. This value is also in terms of each kilogram of target subject's body weight. Tap water usually contains minute traces of chloroform, and a concentration of 0.050 mg/litre is not uncommon in USA water supplies. Thus an adult (weighing 70 kg) who drinks 2 litres per day might well ingest 0.100 milligrams of the contaminant and so be exposed to a lifetime cancer risk of $(0.10 \times 0.006)/70$ or 0.000008. This implies that eight people out of every million are likely to develop cancers from this one cause.

Although the logic of the risk assessment is clear, it is important to stress the assumptions on which it is based—namely:

(a) That chloroform is actually a human carcinogen. The EPA carcinogen potency risk for chloroform was actually derived from animal studies, since no human derived data exists. This obviously casts some doubt on whether the risk assessment is in fact meaningful.
(b) That the dose–response equation, for chloroform, adopted by the EPA is in fact correct. Almost invariably these equations are based on the effects of large doses and the responses those create. At much lower doses, such as those which typify contaminated land, high dose–response relationships

need not hold good, and it is possible (despite the regulatory stance adopted by the EPA) that some "safe" threshold dose could in fact exist.
(c) That the assumed relationship between the doses intaken, and the doses which are retained to adversely affect target individuals, are the same at all dose rates. There is, in principle, no reason to assume that this is the case.
(d) That any individual would be exposed every day of his/her life to the anticipated contamination intake rates.

These assumptions may or may not be justified, and the EPA stance is that an over-cautious attitude is essential for public safety. However, this in no way is any guarantee of accuracy, or of the reality of the quantified risk.

Thus it is apparent that the distinction between USA and European practice is *not* the possession of more accurate and precise data on contamination hazards, *but* on the stress which the EPA gives to the use of a consistent and formal risk assessment methodology. This ensures that sites are judged on the same impartial and relative basis and, in this aspect, USA practice is well in advance of what is normal in Europe.

The EPA system calls for assessors to:

(a) Establish which contaminants and contaminant concentrations exist on a site.
(b) Identify those indicator contaminants, on which the more time-consuming aspects of the risk assessment process will be focused. Usually these indicators will be the more abundant contaminants, as well as those which are thought to pose the greatest potential hazards to human health.
(c) Identify the pathways via which the indicator contaminants could arrive at the target/receptor.
(d) Quantify the contaminant exposures involved and establish the likely dose intakes which will occur.
(e) Quantify the risks involved (as in the chloroform example).
(f) Check whether the consequences of any change in conditions or any extreme events could make risks greater.
(g) Decide which remediation is necessary to reduce the risks to an acceptable level.

The rigour and consistency implied by this process is obviously beyond criticism. What is less defensible is, however, the use of very simplistic exposure models to quantify contaminant exposures (e.g. simple steady-state balance equations can be used to predict off-site contaminant movement via storm run-off). The use of rather arbitrary contaminant hazard ratings, whilst criticised above, is not in itself a difficulty. The values used are as valid as any others which could currently be proposed.

More importantly, it has to be noted that the EPA type risk assessments are long, appear likely to absorb a good deal of scarce technical expertise, are said to be distinctly more expensive than conditions in the UK and Europe seem likely to permit, and focus almost entirely on human health risks. Thus their adoptions in the different European regulatory conditions would not necessarily be appropriate.

Neither the currently utilised qualitative risk assessment method, nor the apparently more indicative EPA quantified risk assessment approach, adequately meet current UK needs for a defensible, logical and relatively inexpensive approach to the whole range of risks which particular contaminated sites might pose.

However, a semi-quantified approach, including elements of both longer established methods, does offer the hope of meeting the defined needs. This is detailed in Chapter 5.

3

Hazard Identification

3.1 INTRODUCTION

It would obviously be futile to attempt any risk evaluation until the scale of hazards, which exists, has been established. In the context of land contamination, this requirement translates to ensuring that site sub-surface conditions have been adequately typified, by an appropriate level of site investigation complemented by a suitably diagnostic chemical analytical strategy.

3.2 SITE INVESTIGATIONS FOR CONTAMINATED LAND

3.2.1 Introduction

No site investigation can ever be certain totally to reveal what lies below ground surface. Even on an undisturbed "greenfield" site—where a regularity of soil and rock layers is more likely than not—only a trivial percentage of a site's volume can ever be examined and sampled. Thus it is unsurprising that technical literature lists so many examples where construction work had to be delayed or re-designed because unpredicted ground conditions were encountered (Institution of Civil Engineers, 1991).

This clearly indicates that site investigation is itself an exercise in risk assessment, and that attempting to entirely uncover the hidden complexities of the sub-surface conditions will usually be unrealisable. Instead what is needed is a high probability that the predicted underground conditions will prove to be accurate.

If it is difficult to predict ground conditions accurately on an undisturbed site, the problem is much worse when contaminated land conditions have to be investigated. This stems quite simply from the decades to centuries of prior industrial uses which contaminated sites have undergone. During this time, changes in plant and building layouts, variations in the industrial processes which took place, and the maze of excavations, infillings and tippings of wastes and surplus materials have usually combined to give very non-uniform near-surface conditions, on at least areas of such sites (Figure 3.1).

These complexities have to be understood, and an adequate sampling and chemical analytical programme (within as low an investigation budget as is achievable) has to confirm which hazards actually do occur.

Attaining this level of understanding is best guaranteed if—instead of the traditional single-event site exploration—investigators utilise a rational multi-phase approach, for example:

(a) preliminary site investigation—desk study,
(b) preliminary site investigation—site zoning,
(c) establishment of initial site zoning model,
(d) main site investigation to prove/revise the site zoning model,
(e) establishment of site zoning model and prediction of contaminant states, and
(f) additional (design) site investigation focused on those details needed to establish whether or not a chosen remediation approach will be effective.

Site zoning is the division of a site into various sub-areas within each of which a particular and different past land use or set of uses is likely to have given rise to a discrete and distinct ground contamination condition.

Whilst the above listing could suggest that this approach must always be expensive, the reverse can in fact be the case. The time available for thought and evaluation between the actual site work phases (i.e. (b), (d) and (f)—above) usually prevents the significance of individual items of information from being overlooked, and so it will be less likely that unpredicted ground conditions will be encountered and affect the planned construction schedule.

3.2.2 Preliminary site investigation—desk study

All too many investigators view site investigation as the physical process of constructing exploratory holes. These individuals probably enjoy the anticipation of what might be revealed by a particular borehole or trial pit, and so tend to charge on to sites to construct as many exploration holes as quickly as possible. Whilst this approach certainly has its interest and excitement, it tends to serve the client's interests very poorly indeed. Instead of each item of information forming an element in a comprehensive and evolving site model, what tends to be produced is a great mass of unrelated points of detail, without any strategy or means of assessing the importance which ought to be ascribed to an individual item of information.

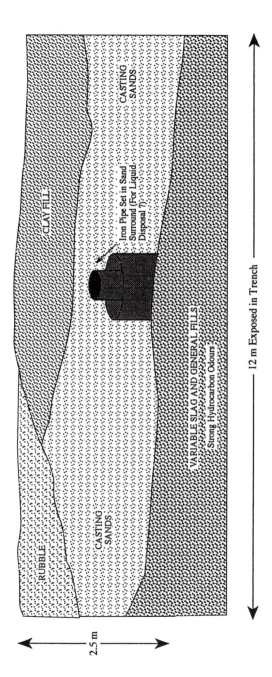

Figure 3.1. Variability on a former metal-working site (West Midlands)

Box 3.1. Main information sources for desk studies

Prior uses of the site

   Past editions of Ordnance Survey (OS) maps
   Past editions of town and local maps
   Past editions of trade directories
   Various generations of aerial photographs
   Waste disposal authority records
   Newspaper archives
   Evidence from past employees

Site sub-surface conditions

   Geological survey maps and memoirs
   Coal and other mining records

Site contamination state

   Local authority environmental health records
   National Rivers Authority water quality monitoring data
   Waste disposal authority records

A more effective approach is to pose and answer three simple questions:

● What is already known about the site?
● What is unknown and *relevant* to the planned re-use and its environmental impact?
● What needs to be known *now*?

A good deal of information will usually be easily and cheaply available from publicly available sources (Box 3.1). Other less easily located detail will also exist (e.g. maps regularly updated by water, gas and electrical service organisations, local historical archives and journals, etc.) and it could be possible to spend very long periods acquiring all the data which exists. This, however, is seldom necessary or even desirable, and it usually is enough to consult the more easily accessed records, and then decide if further desk study will be worth the effort and cost of uncovering harder to locate records.

In the case of the site shown on Figures 3.2 to 3.6, various generations of town plans and the drawings produced by site occupiers were available. These, together with the recollections of past employees, proved to be enough to establish what had occurred on the land. This then permitted 10 discrete "zones" (Figure 3.7) to be identified. Within each of these, ground contamination conditions seemed likely to be essentially similar.

Events on this site were:

(a) *Coal mining* (pre-1887 to the 1920s). Three coal pits (Nos 10, 12 and 14) existed plus a canal arm for coal transportation. The canal's length was

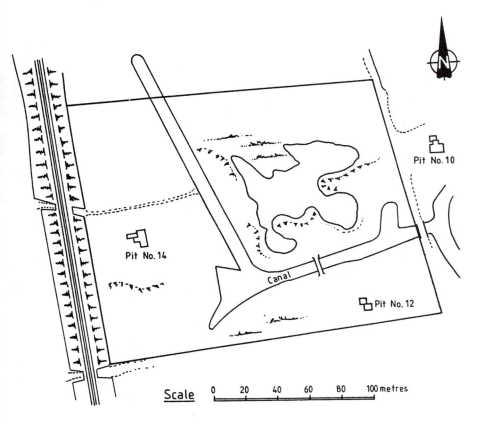

Figure 3.2. 1887 site plan

reduced by 1903, and by the 1920s the canal had been entirely infilled with colliery shale from the spoil heaps then on the site. Site drainage was poor and large ponds existed at different locations. Probably this period did not contribute any signficant contamination hazards, though mine gas emissions could be one result of importance.

(b) *Site unoccupied* (1920s to 1940). No additional contamination hazards postulated.

(c) *Iron foundry* (1940 to 1960s). The large World War II iron foundry on the southern half of the site initially appears to have been devoted to munitions work. During this period an eastern refuse tip (mainly for foundry casting sands and industrial scrap) and a northern (local authority) domestic refuse tip also existed. Site drainage remained poor with a large shallow pond on the northern site boundary. The refuse tips and the spillages in and around the foundry are likely to pose the major contaminant hazards.

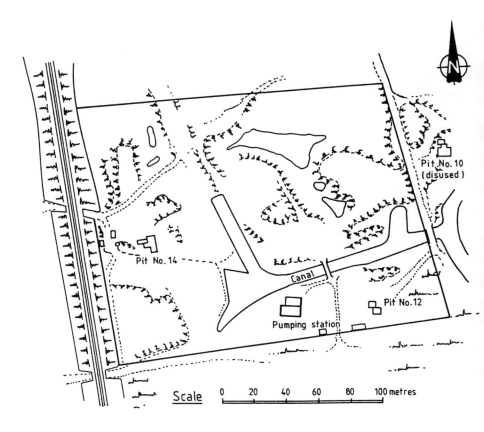

Figure 3.3. 1903 site plan

(d) *Iron foundry—enamelling* (1960s to 1984). The original foundry building remained unchanged, though the activities changed to enamelling cast iron baths, using a hydroscopic and patented mixture of various metallic salts. Additional smaller buildings were added and the two refuse tips were covered and concealed. Site drainage works removed all traces of the former shallow ponds. The use of hydroscopic metal salts for enamelling suggests a significant contaminant hazard, particularly since stores of these salts were spread over the site in the 1984 demolition phase.

On a transparent overlay, it then is possible to mark areas likely to pose particular contamination or other hazards (Figure. 3.7). Areas "A" and "B", on this figure, are the likely areas where refuse tipping might well have occurred, and so where landfill gas hazards and/or heavy metal contamination would be expected. Areas "C", "D" and "E" are where casting sands (likely to be metal

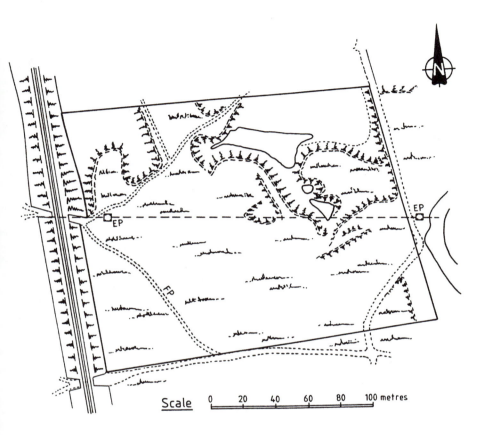

Figure 3.4. 1938 site plan

contaminated) were once stored. The other identified areas are where oil contamination (from the tanks in areas "F" and "G") or where metallic salts (from the patented mixture used in the enamelling process) were spread because of poorly organised demolition work.

Figure 3.7 is, of course, a personal evaluation of available evidence, which does depart in some detail from the earlier (Section 3.2.1) definition of how site zones should be defined. In particular:

● Most investigators would have established a site zone around the whole of the larger iron foundry building. This was felt to be unnecessary, since past employees were able to advise that only the locations where enamelling had taken place (zones "I" and "J") were likely to be sites of contaminated soil. Elsewhere the iron foundry building seemed not to have housed past contaminating activities.

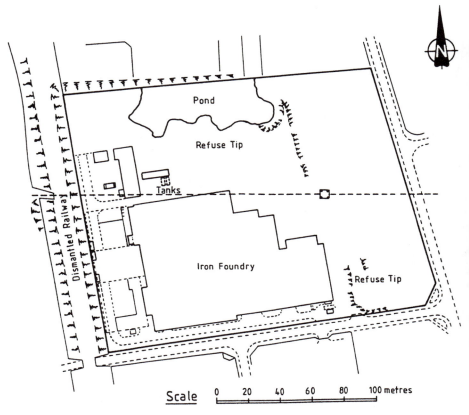

Figure 3.5. 1967 site plan

- The infilled canal arms, also, could have been seen as separate site zones, in which lateral migration of contaminated waters could be occurring. This view, however, seemed very unlikely, given the known facts that the canals had been excavated into Coal Measures Shales and then infilled later with the same materials. Thus these zones were omitted.
- Zones also could have been drawn around each of the former coal mine shafts, as these could still be the locus of mine gas emissions. This was believed to be distinctly unlikely, given the very long time these shafts had been infilled and covered. In fact, this assumption proved to be incorrect, as mine gas emissions were later found to arise from pit No. 12 (see Section 3.2.4).

However, it is not a matter of any real consequence whether or not the Figure 3.7 pattern of zones is correct or incorrect; what is actually important is that a site model has been established and that clear and understood assumptions have been made. As site investigation detail comes to hand a framework exists

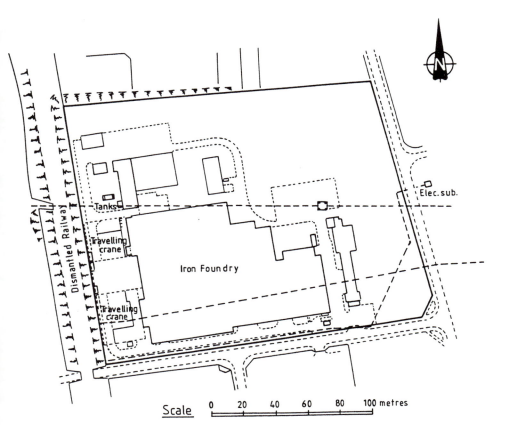

Figure 3.6. Site layout at 1984 closure

against which the relevance of this detail can be assessed. Should this additional information indicate that the site zoning is not an accurate representation of contamination states on the site, then the zoning will—of course—have to be revised. This, however, is not a major task.

Whilst in this case the available plans of the site proved adequate to reveal the sequence of past uses of the land, it has to be cautioned that over-reliance on a single source of information can be misleading. Changes to sites can occur between the production of different plans and so go unrecorded (Box 3.2).

3.2.3 Preliminary site investigation—site zoning

Having completed the desk study, the next stage is to carry out that minimum amount of actual site exploration to confirm the site zoning.

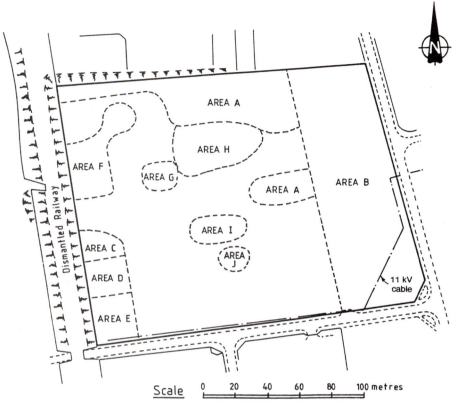

Figure 3.7. Zoning of the site

Box 3.2. The problem of over-reliance on a single data source in desk studies

An executive housing development was planned on a large site adjacent to a long-established village in Central Scotland. The site was visited and proved to be supporting a healthy grass cover established in a clay soil. Conditions appeared to be entirely natural.

Past Ordnance Survey maps (1857, 1897, 1912, 1957 and 1981) all showed the site as a series of fields. No indications of industrial or of non-farming activities were apparent. On this basis, the site was acquired as "greenfield" land.

The maps, however, had failed to record that quarrying of sands and gravels (to 13 m depths) had occurred in 1961 and that the quarry void had then been infilled with "inert" wastes, mainly from road-building works until 1969. The wastes, which were loosely compacted and so offered poor foundation conditions, were also pervasively contaminated by oils and tars.

Whilst the Ordnance Survey maps did not identify either the quarrying or the later tipping, the local waste disposal authority had records of the tipping. Additionally the local Electricity Board's plans showed the extent and depth of the quarrying, since this was adjacent to one of that Board's transmission power lines.

The first phase of this is to:

(a) Walk the site, establish landmarks and identify precisely the boundaries of the postulated site zones. This stage is always best left until the desk study has been completed, since otherwise it could lead to an investigator entering an unsafe area (e.g. in the example given, if one of the known refuse tips had accepted large volumes of asbestos lagging, it would be unwise to walk that area unprotected or to disturb the surface).
(b) Examine the surface of each zone for points of interest, which should then be marked on a large-scale site plan. In the case of this particular site, this led to the boundaries of the two refuse tip zones being slightly redrawn, since waste material could be seen in some shallow excavations and cuttings.
(c) Re-walk each zone looking for any abnormal materials, suface colorations, or abrupt changes in vegetation state which could indicate sub-surface contamination effects. Examine all ditches, drainage lines, etc., for contamination, and check if abnormal odours occur. Use of a spade to expose slightly deeper materials can be helpful.

Since, in our example, no evidence for any major alteration of the already chosen site zones is found, and since the contaminants probably of the greatest significance were thought to be at quite shallow depths (i.e. <2.5 m), trial pitting was seen as likely to be adequate to check whether or not the site zoning had been well based. Trial pits have the advantages of cheapness and also permit an investigator to see what is in the ground, and observe relationships between a particular deposit and those adjacent to it.

This trial pitting was planned to prove a particular hypothesis on the likely distribution of contaminants, as is advocated by the International Standards Organisation (1993) and by the authorities in the Netherlands (Nederlands Normalisatie-Institut, 1990). Thus it was not necessary to follow the density of sampling locations advised by the British Standards Institution (1988). The chosen procedure was to excavate a first trial pit in the centre of each postulated zone and proceed on a 25 m grid from that initial position, until such time as subsoil conditions changed, and it was apparent that a zone boundary had been crossed. The large area of the site not contained within any of the chosen contamination zones was trial pitted, at a lesser density, to ensure that this was underlain by no more than colliery shale material, which had earlier (Figure 3.4) been stockpiled in large colliery spoil heaps. In all some 55 trial pits were excavated within a short investigation period.

Solid samples were obtained from several depths in each trial pit and stored in one-litre screw-topped glass jars. Where oily contamination was apparent in solid samples, representative samples were dispatched without delay for chemical analysis, since there is well-established evidence that losses of volatile components (the ones most likely to pose actual contaminant hazards to future site users) are prone to take place.

This cheap and rapid site investigation exercise proved:

(a) That the site zoning had in fact been quite accurate. Land contamination patterns did indeed follow the zones shown on Figure 3.7. The zones established were as follows:

- landfill areas—zones "A" and "B"
- iron foundry sand stores—zones "C", "D" and "E"
- oil tank and oil spill areas—zones "F" and "G"
- enamelling salt storage area—zone "H"
- enamelling salt usage areas—zones "I" and "J" in the iron foundry

(b) The most obvious contaminant hazards were:

- the two refuse tip areas, where rotted organic material proved to be widespread,
- the areas contaminated by the enamelling salts, which had been spread by the bulldozing of storage vats,
- areas where contaminated foundry casting sands occur as near-surface deposits, and patches where the ground is heavily soaked with spilled oils and oily fluids.

(c) That no near-surface groundwater occurs, presumably because the site's land drainage works are still operating effectively.

3.2.4 Main site investigation

The planned re-use of this example site was for a mixture of domestic housing and light industrial units. Such re-use obviously indicates that the main hazards of concern will be:

(a) Of foundation materials being attacked by aggressive ground conditions. Sulphate, phenol and chloride contamination are thus likely to be of concern, as would be very acidic soils.
(b) Of gases and vapours entering homes and work places. Thus further investigation of the gassing properties *both* of the two refuse tips *and* of the areas around the former colliery shafts is necessary.
(c) Of direct human contact and inhalation/ingestion of metallic contaminants. This concern focuses mainly on metal-contaminated dusts.
(d) Of plants being exposed to phytotoxic ground conditions in gardens.

Additionally, if spilled oils, oily fluids and more soluble solid contaminants have moved down into sandy bands in the underlying Coal Measures rocks, then groundwater pollution could have resulted. If this were the case, then the new landowner could be exposed to legal action (under the 1991 Water Resources Act) and to substantial clean-up costs (Chapter 7, Section 7.1), or to civil action by future home owners if polluted water recontaminated the site's near-surface layers.

Whilst enough actual work had already taken place to obtain representative samples, which (when analysed) could quantify the actual scale of some (i.e. (a), (c) and (d)—above) of these hazards, it was obvious that this was not so in all cases. In particular, further work was needed to establish whether risks from gases and vapours actually occur, and no knowledge at all had been gained on local groundwater qualities and conditions. The main site investigation thus was focused on these two issues.

Gas investigations required an additional series of shallow boreholes, each equipped with a gas-tight cap and gas taps. Gas concentrations and gas pressures where then monitored at weekly intervals for several months to establish that local "hot spots" occurred. These more hazardous areas invariably exhibited very high methane ($\sim 28\%$), high carbon dioxide (~ 8 to 10%) and very depleted oxygen (~ 3 to 1%) concentrations, but were surrounded by other areas, also underlain by refuse tipping, where more normal soil atmospheres occurred.

A similar exercise around the site of colliery pit No. 12 also showed that, periodically, normal soil atmospheres could change and that quite high methane concentrations, under positive gas pressures, could appear. These abnormal methane levels were found to be associated with periods of particularly low atmospheric pressure conditions. This discovery obviously highlights a deficiency in the site zoning (Figure 3.7) exercise which had to be corrected.

Groundwater investigation was limited to a few deeper boreholes around and down the known geological dip (to the east) from the two zones where oily contamination was known, and to confirming that the infilled canal arms were not in fact channelways for pollution migration. This revealed a lack of any groundwater in the upper shaly horizons of the Coal Measures rocks and only very small groundwater collections at significant depth in a single sandier band.

In conclusion, it proved possible to limit the main site investigation, and to focus it only on those aspects which had not been adequately explored in the initial trial pitting exercise. Thus exploration expenditure was less than might otherwise have been the case.

3.2.5 Chemical analytical strategy

Analytical costs are the largest element in contaminated land site investigations, and tend to cost about £100 per sample for the standard packages offered by many laboratories (Box 3.3). With some 200 solid and water samples, from the trial pitting and borehole programmes, a cost of some £20 000 could be anticipated, if all had to be chemically analysed.

However, cost alone is not the point of real significance. Standard analytical packages (Box 3.3) directly relate to the latest ICRCL listing of contaminants

---

Box 3.3. Typical standard analytical package

"Total" elements	*Anions*	*Other determinations*
Arsenic	Chloride	pH
Boron	Sulphate	Phenols
Cadmium	Sulphide	Toluene extractables
Chromium		
Copper		
Lead		
Nickel		
Zinc		

Possible additions
"Coal tars" (by thin-layer chromatography)
Cyanides
"Mineral oils" (by indirect estimation)
Polyaromatic hydrocarbons (PAHs)

(1987) and so would—on the face of it—seem entirely appropriate for contaminated land evaluations. However, this quite obviously is *not* the case. A simple and apparent example is the determination of those substances likely to be harmful (phytotoxic) to plants. Amongst these are the elements copper, nickel and zinc, but do "total" determinations of these three metals actually indicate the risks to plants?

An earlier version of the ICRCL lisings (1983) did identify the "plant-available" concentrations of these metals which could be of concern, but this degree of sensitivity, for some difficult to understand reasons, has been lost in more recent ICRCL guidance (1987). This obviously is unfortunate, particularly since experience shows that the ratio of (say) "plant-available" to "total" zinc is not a fixed percentage, but varies with the chemical and physical form and the origin of the metal. On a site where phytotoxic metals occur in two distinct and different forms (as the relatively insoluble contaminants in the discarded foundry casting sands, and as components in the very mobile and hydroscopic salts used for cast iron bath enamelling) it is obviously more appropriate to specify that "plant-available" phytotoxic contaminant analyses are necessary on samples where the investigation concern is to avoid adverse contaminant effect on plants grown in domestic gardens.

Similarly, the test for the toluene-extractable content is far from satisfactory. This is a general screening test, not intended to quantify any particular contaminant, but to indicate (by the occurrence of unusually high concentration levels) where it is necessary to carry out more specific and expensive tests for (say) coal tars or particular oils. Whilst this appears a reasonable approach, it has to be noted that no standard methodology exists for the toluene-extractable test and—in particular—there are no agreed extraction times or extraction temperatures. Thus the same sample, analysed by different laboratories, can give results which differ by up to a factor of two. Additionally, toluene will also extract innocuous materials (such as the coke dust which is quite prevalent on the example site's surface), and so mislead an investigator.

The whole question of the reliance which an investigator can place in chemical analytical results is, in fact, much less certain than is often believed. Chemical information obviously is a critical component in the evaluation of contaminated land, yet analytical methods are still not standardised and different laboratories do use extraction methods of very different effectivenesses. Concern over this variability in analytical practices has been authoritatively voiced by the Laboratory of the Government Chemist (Anon., 1993a), and indications are that one in ten environmental measurements is likely to be incorrect.

Some investigators appear not to see this lack of standard and appropriate analytical methods as a problem, since they assume that laboratory accreditation, on say the NAMAS criteria of competence, is an adequate indicator of accurate results. (NAMAS is the UK's National Measurement Accreditation Service and is the national standard to implement the requirements of the International Standards Organisation Guide No. 25 ('General requirements for the technical competence of testing laboratories').)

Such accreditation (which actually is awarded for a particular test procedure or procedures and not for all the analytical work an individual laboratory will carry out) is, however, only a guarantee of internal quality control and analytical consistency within that laboratory, and no guarantee at all that a particular result is in fact "correct" or that a particular analytical method was the most suitable one to adopt.

Given the obvious facts, that many analytical techniques are empirical (and measure not the total concentration of a contaminant but *only* that proportion made available by the chosen preparation and extraction method) and that others do not directly measure the contaminant of interest, a prudent investigator should not be persuaded to accept "standard analytical packages". Instead the emphasis should be on selecting direct tests that are appropriate for the particular site conditions, and the hazards that the planned re-use could experience (Table 3.1).

In the case of this particular example site, the zoning that was possible and confirmed allowed the selection of direct analytical tests which actually were appropriate, and it also proved possible to limit very significantly the numbers of samples which had to be analysed. Since site investigation is itself an exercise in risk evaluation, there is no need to analyse every sample which will (for example) underlie domestic gardens for the entire range of phytotoxic contaminants. Analysis of a selected number of such samples will be enough, particularly if these few analyses reveal that "plant-available" contaminants do exist in high concentrations. The prudent assumption is then made that such conditions will occur throughout the areas of the site where plant cover is to be established.

Table 3.1. Analytical strategy based on probable risk situations

Probable risk	Analytical bias required
1. Surface water pollution	Emphasis on material water solubility and mobility (e.g. oils, solvents). Emphasis on ease of water erosion.
2. Groundwater pollution	Emphasis on material water solubility and mobility (e.g. oils and solvents). Emphasis on site permeability and infiltration by rainfall.
3. Air pollution	Emphasis on dust-sized particle chemical condition and on acid soluble results (since the human gut is acidic). Emphasis on gases and vapours.
4. Vapours and gases to on-site users	Emphasis on gases and vapours (*not* only on landfill gases).
5. Attack on materials	Emphasis on water soluble contaminants (particularly sulphates and chlorides), on mobile phenols, and on the acidity of soils and waters.
6. Attack on plants	Emphasis on plant-available contaminants (ADAS methodologies).
7. Attack on site users' health	Emphasis on dust-sized particle chemical condition and on acid soluble results.

3.2.6 Design site investigation

Additional investigation is usually required to prove that a chosen remediation method will be effective in the unique conditions of a particular site.

This, in the context of the example site, included trials:

(a) to prove that the installation of stone-filled vent trenches fully removed landfill gas concerns,

(b) to determine whether a dry separation variation of the soil washing remediation technique would be effective in removing foundry casting sands and their metallic contamination, and leave behind uncontaminated soil,

(c) of methods to collect and sterilise the metallic hydroscopic enamelling salts, which had been spread over large areas of the site during demolition of the iron foundry, and

(d) of methods of degrading the oily contaminated soils and of reducing concentrations of volatile compounds by aeration.

Such additional investigation is particularly necessary if remediation intends to make use of newer reclamation methods, which are process based. These newer methods can often be negated by the presence of other soil contaminants (Anon., 1989) and so have to be proved against the conditions on each individual contaminated site.

3.3 CONCLUSIONS

A logically defensible and multi-phased site investigation is, by far, the soundest and most cost-effective approach to the complexities which typify contaminated sites.

If a site zoning can be postulated from desk studies, and then confirmed or revised as information from the various site investigation phases becomes available, a framework will exist to bring out the actual significance of each item of sub-surface information. This prevents an investigator overlooking the significance of a particular piece of evidence. For example, had a trial pit located to the south-east zone "G" (Figure. 3.7) shown oily liquors oozing into the excavation, this could have indicated that the more northerly of the infilled canal arms was acting as a drainage line, allowing spilled oils from around "G" to contaminate a long linear zone. In such circumstances, a few additional trial pits along the line of the old canal arm would probably have been enough to make the position clear.

It can, of course, be argued that the chosen example site lent itself particularly well to the site zoning approach. Other, less easily divisible sites do occur, and more resemble ill-engineered and very variable waste disposal tips. In such conditions, site zoning will be a less powerful tool and investigators will necessarily have to give greater weight to using regular investigation and sampling grids. However, this is no more than a recognition of reality, and wherever it is possible, zoning a site offers real benefits. Where this is not achievable then a more difficult and expensive investigation task has to be faced.

Site investigation and its associated chemical analyses are carried out primarily to identify where hazards exist and precisely in what form they occur. Until this is done, any attempt to produce a meaningful risk analysis is likely to be less than satisfactory.

4

Pathway and Exposure Assessments

4.1 INTRODUCTION

Once a site's hazards have been established, it remains to identify which pathways exist, via which contamination hazards can affect the likely targets, and whether the scale of the contaminant exposure which targets could suffer can be gauged.

4.2 PATHWAY IDENTIFICATION

Any well-conducted site investigation should indicate which pathways, in the context of a particular site's unique conditions, are likely to be important. The sole cautionary note is that investigators have to keep in mind the likely re-uses to which the site is to be put. Quite obviously differences in site future use do affect the relative importance of different contaminant migration pathways.

Simple illustrations of this are possible from the example site cited in Chapter 3.

If the colliery spoil heaps (which had featured until the World War II redevelopment) had persisted until more recent times, it would obviously have been necessary to excavate and re-grade these heaps to give a more acceptable landform. Such disturbance of fresh colliery spoils implies exposure to oxidation and to rainfall. This can gradually give rise to enhanced concentrations of chlorides, sulphates, various metals, and often increased acidity (down to a pH of perhaps 3.0) in waters draining off or percolating through the disturbed colliery spoils. The processes and exposure durations—whilst complicated by the internal chemistry and buffering capacities—which give rise to these chemical changes are quite well understood (Twardowska *et al.*, 1990) and it is not especially difficult to predict (from leaching tests and by monitoring the changes in small trial embankments) what is likely to occur (Box 4.1).

If such testing had shown that these particular colliery spoils were indeed sensitive to oxidation effects, the consequences for redevelopment and the contaminant migration pathways of importance would vary depending on the planned land uses.

Box 4.1. Variations in the quality of water draining from a colliery shale embankment (after Twardowska and Szczepanska, 1990)

For the first six months of exposure to rainfall, outflow waters were found to be alkaline (pH of 6.5 to 9.0) and typified by declines in the chloride (7000 to 50 mg/litre) and sulphate (8000 to 1000 mg/litre) concentrations.

Later, as the colliery spoil's buffering capacity declined and acidic conditions appeared, outflow waters become very acidic (pH of 2.7) and sulphate concentrations rose significantly (to 6000 mg/litre).

Other colliery spoils with different carbonate contents may/may not develop acidic outflows more rapidly.

A light industrial and warehousing development probably would not be subjected to any significant risk. This can be claimed because:

(a) Such projects are completable in quite short periods. Excavation, re-grading, compaction and installing drainage works in the colliery shales would be a rapid process, taking no more than a few weeks to complete.

(b) With only a limited time for colliery shale exposure to atmospheric conditions, only a minor degree of oxidation would be likely. The effects of this could be predicted and would probably lead to no more than a slight enhancement of ground contamination.

(c) Hardstanding and roadway cover over the site could then proceed quickly, and cut off further access to atmospheric oxygen and moisture. This would prevent any more extreme chemical changes developing.

The likely outcome would be that direct contact (between concrete and other foundation materials and the surface of the underlying compacted colliery spoils) might pose the sole post-construction contamination risk, and this could be adequately prevented by selecting concrete mixes (Building Research Establishment, 1991) or other construction materials which would not be susceptible to the ground conditions.

A similar conclusion could result if the planned re-use were to be town houses and blocks of apartments, set in hardstanding areas.

However, if any sizeable part of the site area had to support a vegetation cover, matters would be unlikely to be so straightforward. Plants are particularly susceptible if rooted into oxidising colliery shales, and either uptake and concentrate contaminants or die off when exposed to conditions where contaminated soil moisture is more prevalent. Also, to ensure that plants can develop and grow, loose textured ground is essential (Dobson and Moffat, 1993) and so air and water would be able to continue to infiltrate into the shales. On-going generation of contamination thus would be likely, and it would not be possible to state that a near-surface (and probably contaminated) groundwater table would not develop. This would imply that the post-construction contaminant risks would include:

(a) the same contact risks to construction materials noted earlier,
(b) risks to plant life and to people or animals who consumed these,
(c) risks to the wider environment as mineralised and possibly acidic waters drained down to local aquifers or entered surface streams, and
(d) contact risks to people, if the vegetated areas were in gardens or allotments.

This case would obviously call for an evaluation of many more contaminant migration pathways than was necessary in the light industrial/warehousing land use.

A similar set of arguments could be advanced if the two refuse tips (which exist on the example site) could not be adequately treated, to remove any future concerns over landfill gas generation, before any buildings were erected.

For a light industrial and warehousing use, risks would arise if these gases could enter enclosed spaces in the industrial buildings. These risks could be reduced to more acceptable levels by installing gas-proof membranes in foundation slabs and ensuring an adequate ventilation of these foundations (Smith, 1993). If this were done, contaminant migration pathways would still exist wherever the gas-proof membranes had to be pierced (by—for example—service pipes and ducts). The maximum acceptable entry of methane could then be quantified on the basis of fractures or cracks of particular size occurring (Box 4.2), and it could be possible (though probably not desirable) to decide to locate industrial premises on such a gassing site. One factor which would make it easier to accept this conclusion would be the assurance provided by knowing that the organisation which owned a large industrial unit would have a central management, which could be fully advised of the risks, and of the need to avoid any future interference with the gas-proofing measures.

If, however, detached domestic houses, each with its own private garden, were to be built, no gas-proofing solution would be easily accepted. Pathways for gas migration could occur in the garden areas, where home owners might erect greenhouses or home extensions, which could then become gas collection points. Additionally, plant life in gardens could be adversely affected. Methane is known to be able to displace oxygen from plant root zones and so effectively asphyxiate the plants. Worse effects would also be possible if carbon dioxide moved up into the rooting layers or if the methane were oxidised to carbon dioxide in near-surface soils. Carbon dioxide is known (Leone *et al.*, 1979) to be directly toxic to plant life at particular concentrations. Finally, gas-proofing domestic dwellings is not a fail-safe solution, since home owners can compromise or disrupt gas-proofing measures by later home modifications. Thus, if the gas risks from the refuse tips could not be fully resolved, and if any relatively large generation of landfill gases were feasible, it would be necessary to consider a good many more contaminant migration pathways than was necessary had the

Box 4.2. Evaluation of acceptable methane entry rates into enclosed spaces (after Smith, 1991)

Identify the smallest enclosed space at risk (say a broom cupboard containing an electrical switch box)

Volume of broom cupboard (say)   2.5 m^3

Methane entry rate to achieve lower explosive level (LEL) of 5% methane in air   125 litres

Likely maximum period when broom cupboard remains undisturbed and unventilated (in—say—annual holidays)   15 days

Methane hourly entry rate to achieve LEL in this period   0.347 litres/hour

Allowing for a factor of safety of 5 (i.e. assume that LEL concentrations should not occur on any of the 15 days) reduces this maximum allowable entry rate of methane to   0.069 litres/hour

Accept a typical small building has a plan area of (say) 40 m^2 and assume that settlement below the structure could occur. Thus the maximum allowable methane entry would be   0.0017 litres per hour per m^2

If the building's foundation slab had been gas-proofed, concern would focus on cracks around service ducts. If the total area of such cracks proved to be 750 mm^2, the safe entry level of methane in the specified conditions would be only   0.0016 ml per hour per m^2 covered by the building

land been used for light industrial purposes. Hopefully any investigator would find these pathways of such concern, that his or her advice would be against proceeding with a domestic development. Too many cases, where such developments have not proved safe, are known to encourage any more.

Pathway identification is not difficult if an adequate knowledge of site sub-surface conditions exists and if the planned land re-uses are understood.

Much of the exercise consists of identifying geological and near-surface horizons and deciding which migration routes these offer to more mobile and soluble contaminants. Additionally, it is essential to note that artificial discontinuities (such as sewers, road sub-bases, etc.) can give rise to localised high permeability pathways.

It is, however, important to avoid casual decisions (e.g. Box 2.1, Site C) or any inference that a geological description implies particular physical properties and migration possibilities. When considering whether landfill gases could migrate laterally to affect land some distance from a gas generation "hot spot", some investigators could take the view that the presence of clay layers surrounding the generation source would adequately prevent lateral gas movements. This view rests on the assumption that clays are always "impermeable". Whilst this could be the case (though not invariably) when the migration of water is being considered, it obviously is an improper assumption when the fluids of concern are gases. Gas impermeability of natural materials is very much a function of whether their pores are water filled or dry. A fully water-saturated clay is indeed likely to be a very effective barrier to lateral movement of any large volumes of landfill gases. However, when partly desiccated, the same clay can prove to be extremely permeable to gases.

It also will always be essential to identify worse-case conditions which could enhance pathways. Groundwater levels will vary on an annual basis. So if site investigation took place in mid-summer (when groundwater is often at its greatest depth in a site) conditions might well seem very unlikely to permit upward migration of contaminated groundwater to affect near-surface foundations and plant rooting zones. The same situation, however, need not occur in early spring, and it could be possible then that foundations could lie only a few centimetres above contaminated groundwater and so be exposed to contaminated soil moisture.

4.3 EXPOSURE ASSESSMENT

Whilst pathway identification is usually possible without undue difficulty, the same cannot be said for the assessment of those exposure doses a target could suffer.

The above comment may seem strange, since US practice is able routinely to gauge human health risk exposure doses (Chapter 2, Section 2.4). However, the need in US practice to assess the weights of particular contaminants which might be eaten or ingested is an unusually simple exercise (Box 4.3). All that is needed are the assumptions detailed in Chapter 2 (Section 2.4). For the other

---

Box 4.3. Typical human health risk exposure assessment (assumptions as listed in Chapter 2, Section 2.4)

Concern could be voiced that eating garden crops grown in contaminated soil could pose direct health risks. If so, exposure assessment could be gauged, i.e.:

Maximum concentration of a particular contaminant in garden produce (dry weight)   say 10 mg/kg
Likely maximum intake of garden produce by individuals (proved by survey)   say 500 g/day
Peak daily contaminant intake   5 mg
Period of crop availability   30 days/year
Peak annual contaminant intake   150 mg

If this dose equalled or exceeded national or international safety levels, then a health risk could be anticipated and remedial action commenced.

risks which contaminated sites could present, matters are more complicated and meaningful exposure doses will be very difficult or even impossible to define.

An apparently very simple exercise would be to judge how far run-off from a contaminated site could adversely affect the quality of an adjacent and clean river. As the maximum allowable contaminant concentrations in river water, used for public supply, are known (European Communities, 1975), it might seem relatively straightforward to determine the total weight of any chosen contaminant in the river just upstream of the contaminated site, and the total weight of that contaminant added by run-off from the site in wetter weather, and so judge the pollution impact.

This approach would indicate the use of a steady-state evaluation formula of the following type:

$$c^I Q + c^{II} q = c^{III}(Q + q) \qquad \text{(equation 4.1)}$$

where

c^I is the concentration (mg/litre) of that contaminant in the river water, just upstream of the site

Q is the flow rate (m^3/s) of the river, just upstream of the site

c^{II} is the concentration (mg/litre) of the contaminant in waters running off the site's surface

q is the rate of this run-off (m^3/s)

c^{III} is the concentration (mg/litre) of the contaminant immediately downstream of the site

However, even a brief consideration reveals that none of the terms in equation 4.1 has a fixed and absolute value. River flow rates (Q) vary with weather conditions and often by a factor of 1000 or more. As the flow rates change, so also do the concentrations of particular dissolved substances (c^I). Likewise the rate of run-off (q) from a site surface changes with rainfall rate, rainfall duration and season (as the state of surface vegetation impedes or encourages run-off). Even for a specified rainstorm, run-off will not be a constant value, but will gradually increase to a peak level and then fall off, over quite a long time. The quality of this run-off (c^{III}) will also be a variable, affected not only by rainfall factors, but also by variations in the solubility of contaminants on or near the site surface and whether or not run-off carries suspended particles of contaminated soil.

It could, of course, be argued that these variabilities could be resolved by site studies. However, this would involve long-term effort and costs, and would usually not be realistic in the time contexts in which risk evaluations have to be produced.

More complicated would be attempting to quantify the contaminant exposure doses to which a vegetation cover could be exposed. It would not be difficult to determine the depth and quality of a contaminated groundwater in a site, and if this were relatively close to plant root zones (i.e. nor more than 3.0 m below) form the view that this groundwater could be abstracted by plant roots and possibly prove phytotoxic. Contaminant concentrations that are likely to pose phytotoxicity are established in national guidelines (e.g. ICRCL, 1987). However, it cannot be assumed that any particular groundwater quality would remain unchanged as moisture from it moved up through a less than fully water-saturated soil profile. A variety of physical and chemical effects (e.g. cation exchange) could quite seriously alter the moisture quality. Establishing the significance of these effects might be possible, though not within any acceptable time durations.

When the additional filtration abilities of particular plant species' roots, which can separate out cleaner moisture from some unwanted mineralisation are taken into account, it obviously would not be possible to produce any defensible contaminant dose assessment in many circumstances.

A number of risk situations arise from land contamination:

(a) risks of surface water pollution,
(b) risks of groundwater pollution,
(c) risks of widescale air pollution,
(d) risks of gases and vapours entering enclosed spaces,
(e) risks to construction materials,
(f) risks to plant life,
(g) risks to human health via contact, ingestion or inhalation.

For all of these, no particular difficulty should be encountered in identifying the pathways along which contaminant migration could occur. However, quantifying the scale of migration along most pathways will inevitably be difficult or practically impossible. Conditions simply will be too complex and too site-specific to yield simple answers, within any limited assessment time.

Thus, despite the well-known USA practice of quantifying contaminant dose exposures for human health risks, it simply is not justifiable to include this level of apparent precision in the evaluation of the majority of risk situations.

The proposed risk assessment methodology (Chapter 5) thus avoids any role for exposure dose assessments. Contaminant hazards are identified and the pathways significant to the various risk situations are highlighted. The cautious assumption is then made that where the hazard → pathway → target chain is complete, a risk will exist.

5

The Semi-Quantified Risk Assessment Approach

Current inadequacies in national guidelines and standards (Chapter 1, Section 1.3), and the scientific uncertainties over the fates and interactions of contaminants as they move through the complexities of air–soil–water environments are such that a fully quantified environment risk assessment approach cannot yet be advocated.

Equally, it is apparent that qualitative risk assessments are of distinctly limited benefit and far less convincing than is needed to protect clients' interests. The opportunities for individual assessors to adopt very personal emphases are simply too great for any acceptably consistent conclusions (Chapter 2, Section 2.3).

A simple semi-quantified assessment method is proposed. This offers the rigour of the quantified approach, whilst still retaining the flexibility needed to overcome scientific information deficiencies. In this method, assessor freedom of choice is deliberately restricted, gaps in the necessary site information are highlighted and penalised, and assessors must inform colleagues, and the individual in charge of the assessment group, of any abnormal information or conditions. This permits an agreed evolution of the assessment approach, as experience proves necessary.

Use is necessarily made of suites of contaminants and of trigger contaminant concentrations to identify potential land contamination hazards. These have not been based on any new theoretical or research findings, but instead are the parameters and values which experience suggests will be practically appropriate to use. If later studies indicate that these have to be revised, no especial difficulty should result. The locations, where contaminant suites and critical concentrations are made use of, are clearly indicated. So risk judgements should be easily revised, as and when additional research on the fates of contaminants in the complexity and variability of soil–air–water environments becomes available.

Most importantly, care has been taken to make the method as simple, easy and cheap to use as possible. Financial restraints are a major reason why so few European risk assessments have been carried beyond initial desk study stages.

This seems to be a result of a widespread lack of belief that the benefits of fully completing risk assessments will justify the necessary extra costs. Counteracting this view, by producing inexpensive risk assessments, can only be in the best interest of the vendors, buyers and occupiers of reclaimed land.

A total of seven environmental risk situations is considered:

Group 1   (1) Risks of polluting surface waters
            (2) Risks of polluting groundwaters
            (3) Risks of producing area-wide air pollution
Group II (4) Risks of gases and vapours entering dwellings and
                structures
            (5) Risks of attack on construction materials
            (6) Risks to plant populations
            (7) Risks to human health by contaminant contact, ingestion
                or inhalation

The Group I risks are those of interest to the control bodies charged with minimising wider environmental degradation. The powers given by the Environment Protection Act 1990, and the Water Resources Act 1991, are such that creating any of these wider environmental risks is likely to attract particularly expensive clean-up costs, legal enforcement actions, or even fines and penal sentences in more extreme cases. Civil actions could also arise, particularly once a statutory control body's actions had alerted local inhabitants. A prudent developer obviously would prefer not to acquire any site which could be a locus of such problems.

The Group II risks are those of direct concern to individuals or companies owning or occupying land, which formerly was contaminated. Whilst concerns that legal actions might occur will still exist, if reclamation has not been adequate to remove these four risk categories, the more dominant worries are those of financial loss, failure to be able to sell reclaimed sites, adverse publicity, and the immorality of exposing individuals and families to unacceptable and unsafe situations.

These seven risk categories adequately cover the risks of concern to the various parties (Chapter 1, Section 1.5) interested in the re-use of a formerly contaminated site. It could, of course, be argued that a fuller range of risks should be addressed. Some workers (Ferguson and Denner, 1993) have suggested that the human health risks from ingesting and inhaling contaminants should be treated separately, and that finer distinctions should be drawn between (for example):

(a) direct ingestion of contaminated soil (particularly by young children playing on unpaved areas),
(b) ingestion of soil on inadequately washed vegetables, grown in contaminated soil, and

(c) ingestion of contaminated dusts, which can reach higher concentration levels within inadequately ventilated homes than occurs outside.

Whilst it is not disputed that these arguments have force, increasing the complexity of risk analysis beyond the necessary minimum level is difficult to justify in the current absence of specific and detailed scientific information (Chapter 1, Section 1.3).

For each of the seven liability situations a three-part risk assessment approach is adopted:

Part [A] The potential for future risk is identified by the desk study approach familiar to most assessors (Chapter 3, Section 3.2.2). The sole differences from normal practice are the limitations imposed on assessors, and the requirement that any decision to terminate the risk assessment process (immediately after this initial phase) must be justified. No numerical assessment values are allocated, since this stage is essentially a screening exercise designed to determine whether or not further assessment is needed

Part [B] The quantification of the probable magnitude of a future risk. In this, information from site investigation is used, and any limitations of the information are highlighted and penalised. Assessor choices remain constrained, and colleagues, and the individual in charge of the assessment team, have to be advised if abnormal data (which fails to fit into the standardised method) are encountered. Scoring of liabilities (as negative values) allows the relative magnitudes of the different risk situations to be established. This, in turn, indicates the particular emphases which should be observed where contaminated land remediation methods are being selected, and prevents the importance of a specific risk from being overlooked.

Part [C] The focus here is on how far site remediation has removed/reduced risks. Results (as positive scores) are subtracted from those earlier obtained in Part [B] to give as precise as possible indicators of the scale of remnant liabilities. Particular emphasis is given to proof being available that remediation has, in fact, been effective.

The above three-part process fits well with the usual practical requirements, which tend to fall into one or other of the following categories:

● Pre-purchase decisions when an unreclaimed and contaminated site is offered for sale. Prudent purchasers will want to know if the potential liabilities will be acceptable and be given an indication of the likely costs of site investigation and final remediation. Any site *particularly* likely to give rise to Group I risks

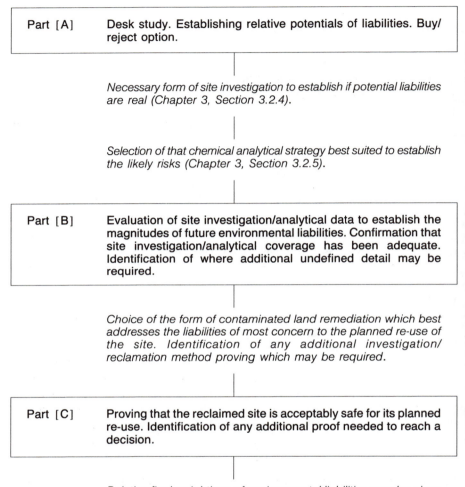

| Part [A] | Desk study. Establishing relative potentials of liabilities. Buy/ reject option. |

Necessary form of site investigation to establish if potential liabilities are real (Chapter 3, Section 3.2.4).

Selection of that chemical analytical strategy best suited to establish the likely risks (Chapter 3, Section 3.2.5).

| Part [B] | Evaluation of site investigation/analytical data to establish the magnitudes of future environmental liabilities. Confirmation that site investigation/analytical coverage has been adequate. Identification of where additional undefined detail may be required. |

Choice of the form of contaminated land remediation which best addresses the liabilities of most concern to the planned re-use of the site. Identification of any additional investigation/ reclamation method proving which may be required.

| Part [C] | Proving that the reclaimed site is acceptably safe for its planned re-use. Identification of any additional proof needed to reach a decision. |

Relative final weightings of environmental liabilities may be given standardised cost values. This will be necessary if the assessment process has been carried out to establish if insurance cover can be given and the premiums which should be charged.

Figure 5.1. Risk assessment flow chart

would not usually be acquired since the remediation costs would probably prove to be excessive.

● Deciding which remediation method best fits the remediation budget, the planned land use, and the liabilities which are of most concern. Failure to conduct a Part [B] assessment can leave the status of a particular environmental risk entirely unclear (Appendix I—example Site No. 2, where

a lack of groundwater quality information caused remnant doubts to persist over the reclamation effectiveness).
- Proving that a reclamation has fully satisfied its aims and that the site is acceptably safe for its projected re-use. Currently several major funding bodies require such detailed assurances before agreeing to loan finances to a potential purchaser.

A flow chart of the assessment method makes the relationship of these three stages clearer and indicates the essential links with various stages and types of site investigation (Figure 5.1).

5.1 DEVELOPMENT OF A RISK ASSESSMENT APPROACH

The need for consistent and credible risk assessments of contaminated land became obvious to the author in the late 1980s. At that time, the UK property market was at the peak of its activity, and hardly a week would go past without requests for advice on the purchase of one or more areas of contaminated land. Since potential buyers had received copies of all available site investigations (quite often, indeed, these would be three or more investigation reports for a single site), large numbers of risk assessment statements necessarily had to be read and judged. After a short time, the errors and deficiencies (see Box 2.1) in these assessments became so obvious, that confidence in qualitative risk evaluations vanished.

This led to the initial attempts to devise a more defensible, semi-quantified approach. Full documentation on 22 sites (where the land had been investigated, reclaimed, and then re-examined in detail by later purchasers) was the initial data base which was used.

Over the next few years, as more information became available, the assessment approach detailed in Chapters 6 to 12 was refined and tested. Almost 200 additional site records were used to ensure that a full range of hydrological and hydrogeological conditions were included in this testing, and care was taken to test sites with a wide diversity of prior industrial histories:

- tipping sites
- oil and tar refineries
- oil storage sites
- disused sewage works
- chemical and dyestuff factories
- printing works
- power station sites
- derelict gas works

Details of three relatively typical examples of these sites are included in Appendix I to permit readers to use the assessment methodology.

PART II

ASSESSING THE MORE IMPORTANT RISKS

Standards and criteria used

Specific suites and concentrations of contaminants are used in Chapters 6 to 12 to judge the risks from contaminated land. It should be uncontentious that these various criteria differ; obviously it would not be tenable to argue that any one list of contaminants and concentrations would be diagnostic for all risk categories. The contaminant conditions likely to harm plant populations (for example) are simply not those which could degrade construction materials or cause widespread air pollution.

What, however, is likely to cause disquiet is that the contaminant suites and concentrations used are *not* those given in official guidance. Indeed, the listed criteria are often less severe than are the current ICRCL threshold values. This arises from the approach adopted in devising appropriate judgemental standards, i.e.:

- Some standards (those for assessing the risks of polluting either surface water or groundwater and the risks of possible harm to plant populations) have been *empirically derived*. Several hundred site investigation and land reclamation reports, covering a very wide range of contaminants and geographical conditions, have been analysed to produce these criteria.
- Other standards (for judging if risks to human health or air quality will be significant) rely heavily on health and safety *occupational exposure limits*, in addition to the site investigation and reclamation data base.
- The remainder of the standards (for assessing risks from gases and vapours to home owners/occupiers and of materials' degradation) depend on published research findings as well as the available site investigation and reclamation information.

The risk assessment approach which is detailed in Chapters 6 to 12 does not depend on the use of these particular standards and criteria, though it is strongly believed that these are practically diagnostic. Other workers with different experience of contaminated land may choose to refine and develop the criteria, and this would be a positive trend. Others will prefer to utilise threshold values given in official guidance (provided—of course—that the threshold values chosen are indeed relevant for the particular risk being assessed *and* for the political and legislative realities).

No matter which course is adopted, the particular standards and criteria detailed in Chapters 6 to 12 *should not* be utilised out of their specified context.

6

Assessing the Risks of
Surface Water Pollution

6.1 THE IMPORTANCE OF
SURFACE WATER POLLUTION LIABILITIES

Pollution of surface waters by contaminated land is far from uncommon. Although the most extensive instances are usually caused by large-scale industrial activity and waste disposal—such as the construction of colliery waste heaps (Sleeman, 1990), the spoils created by metal mining and refining (Parry and Bell, 1987a) and landfills—a very large range of past contaminative uses, also can give rise to adverse effects on surface water, even from quite small areas of land (Flavin and Harris, 1991).

The main concern for a developer is that it is a criminal offence to cause pollution of surface waters (Water Resources Act 1991). Thus if a site is shown to be the locus of surface water quality degradation, the landowners and occupiers can, at best, be required to take whatever remedial action the National Rivers Authority, or the equivalent national control body, sees as essential. Failure to comply with such demands can lead to criminal prosecution (with the prospects of punitive fines and even—in worse cases—terms of penal confinement for individuals and company officers who are found guilty) *and* the repayment of whatever costs the statutory control body has itself expended to rectify that water pollution problem.

With consequences of this scale, any prudent developer will either confirm that a site is unlikely to cause surface water quality deterioration, or that any potential for this can be removed at acceptable cost, before acquiring the site. It may not be enough to see the pollution arising from a site as insignificant, simply because local watercourses are already highly contaminated. Later cleaning-up of other pollution sources could occur, and so improve a stream that the effects of the particular site could become apparent at some later time.

Surface water pollution can of course be caused by:

- dissolved contaminants and free liquids leaking into streams, and
- suspended particles of solid contaminants being carried off a site's surface in times of high rainfall,

and these two causes do create some complexity in assessments.

However, with care it usually is not difficult to determine whether a particular site will expose a future owner to this liability. The three-part assessment process to achieve this is detailed in Section 6.2.

6.2 SURFACE WATER POLLUTION RISK ASSESSMENT

Main Concern

The occurrence of surface water pollution which may attract the attention of the National Rivers Authority (or equivalent national control bodies).

This concern will exist if topographic and/or exposure conditions encourage run-off, or if the discharge of polluted groundwater to surface water can occur.

Should a particular suite of contaminants (List 6.1) exist in mobile/water soluble forms, the concern will inevitably be greater.

[A]   Potential for future liability

Tick appropriate box

1. Is the site located adjacent to a surface water body, or do ditches from the site drain to a surface water body?

   (a) Yes. Fronts directly on s.w. body. ☐

   (b) Yes. Within 300 m of s.w. body and/or ditches drain directly to s.w. body. ☐

   (c) No. Nearest s.w. body > 300 m distant and no ditches drain to it. ☐

2. Would site topography encourage run-off to a surface water body? (within 300 m of the site boundary)

   (a) Yes. Site falls steeply (> 1 in 200) to s.w. body. ☐

   (b) Yes. Site is essentially level but is exposed to main storm direction(s) and flooding is likely. ☐

   (c) No. Site slopes away from s.w. body (> 1 in 200), or the flat topography would not be prone to flooding risk. ☐

3. Would any run-off flow easily to the surface water body?

   (a) Yes. Barriers and/or control works do not exist to stop this. ☐

   (b) No. Barriers and/or specific control works do exist (specify details below) and appear to be effective. ☐

Q3. Details of any barriers/control works (continue on separate sheet).

. . . .

. . . .

. . . .

4. Does the site surface appear to have the potential to produce polluting run-off? (refer to List 6.1)

(a) *Yes. Surface materials loose and easily eroded and probably contaminated.* ☐

(b) *Yes. Contaminated materials (sludges/lagoons/landfill/ industrial wastes) occur on site surfaces.* ☐

(c) *No. Site surface well compacted and/or surfaced with non-erodible cover/site appears to lack any near-surface contaminants (List 6.1).* ☐

5. Does the site appear to be creating surface water pollution currently?

(a) *Yes. Polluting seepages visible and can be seen entering the s.w. body.* ☐

(b) *Possibly. Seepages visible on site (ditches/trenches) and could migrate to s.w. body.* ☐

(c) *No. No signs of any adverse effect on s.w. body, or any mobile pollution in ditches or trenches.* ☐

6. Is the quality of any nearby surface water body (likely to be affected by the site; i.e. within 300 m upstream of site) currently high?

(a) *Yes. Waters clear. Aquatic life obvious.* ☐

(b) *Yes. No signs of pollution. Water clear.* ☐

(c) *No. Waters turbid/discoloured/ lacking obvious aquatic life, upstream of site.* ☐

7. Does NRA (or equivalent control bodies) zoning of any surface water, within 300 m, show the water quality to be high?

(a) *Yes. Detail water quality data below.* ☐

(b) *While quality currently is not high, improvement schemes are planned. List details below.* ☐

(c) *No. Water quality currently is poor and likely to remain so. List details below.* ☐

Q7. Control body quality data/planned improvements (**continue on separate sheet**).

. . . .

. . . .

. . . .

8. Does a control body monitor   *(a)   Yes.*   ☐
   surface quality conditions
   within 300 m of the site (either *(b)   No.*   ☐
   upstream or downstream)?

9. Is there any evidence to   *(a)   Yes. See Section 7.3 analysis.*   ☐
   suggest that a near-surface
   polluted groundwater could   *(b)   No. No reason to postulate polluted*
   exist in the site, and that this   *groundwaters being able to enter*
   could flow into nearby surface   *s.w. body.*   ☐
   waters? (confirm by reference
   to groundwater pollution risk
   analysis—(Chapter 7))

● ● ●

If the answers to all the above questions are "No", terminate surface water pollution
liability assessment at this point, and accept that the liability risk is likely to be
negligible.
For other results, summarise the overall evaluation below. Justify clearly any decision
to terminate surface water pollution liability assessment.

. . . .
. . . .
. . . .
. . . .

LIST 6.1

Substances (expressed mainly as solid concentrations) which may give rise to surface water pollution

Group A

Very acidic compounds   (pH < 5)
Very alkaline compounds   (pH > 9)

Note: pH changes which may be predictable (e.g. via oxidation if site disturbance occurs).

Group B

Arsenic salts	> 40 mg/kg
Boron (water soluble)	> 3 mg/kg
Cadmium salts	> 15 mg/kg
Chromium salts (as Cr VI)	> 30 mg/kg
Copper salts	> 300 mg/kg
Lead salts	> 1500 mg/kg
Mercury salts	> 20 mg/kg
Selenium salts	> 6 mg/kg
Zinc salts	> 500 mg/kg

Note: the very variable solubilities of different compounds of the same metals. High solid concentrations need *not* indicate any water pollution risk. Direct or indirect solubility proof is invariably essential for risk scoring (Q B2).

Group C

Sulphates	> 10 000 mg/kg
Sulphides	> 1 000 mg/kg

Note: the very variable solubilities of different sulphates and sulphides and the need for solubility proof in risk scoring.

Group D

Free oils or oil residues	if visibly present (> 1000 mg/kg)
Landfill leachates	if visibly present
Mobile tars/tarry liquors	if visibly present
Organic solvents	if noticeably present
Cyanides	> 500 mg/kg
Phenols	> 200 mg/kg
PAHs	> 5000 mg/kg

Note: the mobility of this group poses a more definite water pollution risk than is the case with other groups. Thus greater weight must be given to the occurrence of excessive concentrations of substances in this group.

Group E

Other site-specific compounds or wastes which are known to be water soluble:

e.g. DDT
     Pesticides and residues   > 20 mg/kg

Phosphates   >20 mg/kg
Timber preservative   >20 mg/kg
    fluids etc.

Note: local site history/use should be employed to indicate which site-specific compounds are of interest.

Scoring Options	*SCORE*
1. Group D substances and those of other groups present in excessive concentrations and throughout much of the site. All, or most, contaminants prove to be leachable/mobile to a significant extent.	50
2. "Hot spots" of at least three groups of contaminants are present at excessive concentrations. Leachability/mobility is only partial.	30
3. "Hot spots" of at least three groups of contaminants are present. Leachability/mobility is relatively slight, and has not led to pollution of near-surface groundwater.	20
4. Some individual contaminants occur at highish concentrations. Leachability/mobility proves to be nearly negligible.	10
5. Some individual contaminants do occur, but either at concentrations below those cited on List 6.1 and/or leachability/mobility proves to be essentially negligible.	0

[B]   Probable magnitude of a future liability

Score (as negative values)

1. Do List 6.1 substances exist in the upper 2 m of the site and at high concentrations?	*(a)* *Yes.*	10
	(b) *To a limited extent.*	5
	(c) *No.*	0
2. If List 6.1 substances exist, at high concentrations, will these be in mobile and/or water soluble forms? (see List 6.1 for scoring options)	*(a)* *Yes. Specify evidence below.*	50–30
	(b) *To a limited extent. Specify evidence below.*	30–20
	(c) *No. Specify evidence below.*	10–0

Q2. Mobility/solubility evidence.

....
....
....

*Score
(as negative
values)*

3. Would local conditions permit polluted run-off to enter surface water? (include data noted in Q A1, A2, A3)

    (a) Yes.   **10**

    (b) *To a limited extent.*   **5**

    (c) *No.*   **0**

Note: distinguish whether pre- or post-reclamation conditions are being judged.

4. Would local conditions permit a polluted groundwater to flow into surface water? (include data noted in Q A9)

    (a) *Yes. Specify evidence below.*   **10**

    (b) *The suggestion is feasible, but unlikely. Specify evidence below.*   **5**

    (c) *No. Specify evidence below.*   **0**

Q4. Evidence of polluted groundwater able to enter surface water.

....
....
....

---

5. If polluted run-off or polluted groundwater could enter surface water, would this be significant? (include data noted in Q A6, A7, A8)

    (a) Yes.   **10**

    (b) *To a limited extent.*   **5**

    (c) *No. Surface water so degraded that site effects would not be significant.*   **0**

6. Is the Site Investigation data adequate for the above questions (1, 2 and 4) to be answered with confidence?

    (a) Yes.   **0**

    (b) *No. Specify particular data deficiencies below.*   **10**

Q6. Information deficiencies.

....
....
....

Advise Environment Division Director if risk evaluation has been biased because of data inadequacies.

Overall Score   % _____

[C] Risk reduction achieved by site remediation

Score
(as positive
values)

1. Have the remediation works
properly focused on
preventing/reducing surface
water pollution risks?

    (a) *Yes.*   10

    (b) *Not entirely. Specify data below.*   5

    (c) *No. Specify data below.*   0

Note: If probable sources of surface water pollution (if this is feasible) have not been properly targeted, then remediation benefits are unlikely to be adequate.

Q1. Failures/deficiencies in remediation plan with respect to surface water pollution risks.

. . . .
. . . .
. . . .

Score only one of the remediation solutions—Q2 to Q5.

2.(a) Have *all* mobile/soluble List
6.1 substances been
removed from the top 2 m
of the site, or reduced to
below the critical
concentrations? (see List
6.1)

    (a) *Yes.*   70

    (b) *Not entirely. Specify data below.*   35

    (c) *No. Specify data below.*   0

Q2.(a)

. . . .
. . . .
. . . .

(b) If all mobile/soluble
substances appear to have
been removed (or reduced
to below the critical
concentrations), does
quality control or post-
remediation monitoring
proof of this exist?

    (a) *Yes. Specify evidence below.*   20

    (b) *Not conclusively. Specify evidence below.*   10

    (c) *No. Specify evidence deficiencies below.*   0

Q2.(b) Quality control/monitoring proof.

. . . .
. . . .
. . . .

3.(a) If the mobile/soluble List 6.1 substances have been left on site, do works exist to encapsulate and/or immobilise these?	*(a) Yes. Specify details below.*	**50**
	(b) Not entirely. Specify details below.	**25**
	(c) No. Remediation was not adequately designed to encapsulate/immobilise the substances. Specify deficiencies below.	**0**

Q3(a). Adequacy of encapsulation/immobilisation.

. . . .
. . . .
. . . .

(b) If encapsulation/immobilisation has been carried out, does quality control and/or post-remediation monitoring proof of effectiveness exist?	*(a) Yes. Specify details below.*	**20**
	(b) Not conclusive. Specify details below.	**10**
	(c) No. Specify deficiencies below.	**0**

Q3(b). Quality control/monitoring proof.

. . . .
. . . .
. . . .

4.(a) If the List 6.1 substances of concern have neither been removed from site, nor encapsulated/immobilised, have any "passive" protection works (e.g. cut-off systems for surface run-off or topographic improvements) been included as a solution?

(a) *Yes. Specify details below.* 25

(b) *To an incomplete extent. Specify details below.* 15

(c) *No.* 0

Q4(a). Details of "passive" protection works.

....

....

....

(b) If any such "passive" protection has been included as a factor, does it appear likely to be effective? (note probable durability)

(a) *Yes. Detail reasons below.* 20

(b) *Only to a partial extent. Detail reasons below.* 10

(c) *No. Detail reasons below.* 0

Q4(b). Likely effectiveness of "passive" works.

....

....

....

5.(a) If the remediation consisted of a particular mix of removal/ encapsulation and immobilisation plus "passive" protection, does this appear likely to be effective?

(a) *Yes. Details and reasons below.* 50

(b) *Not entirely. Details and reasons below.* 25

(c) *No. Details and reasons below.* 0

Q5(a). Mix of remediation techniques and likely effectiveness.

....

....

....

Score
(as positive values)

(b) In any such case, is there quality control and/or post-remediation monitoring to prove the effectiveness of the remediation?

(a) *Yes. Specify details below.*   20

(b) *Evidence not conclusive. Specify details below.*   10

(c) *No. Specify details below.*   0

Q5(b). Quality control/monitoring proof.

. . . .

. . . .

. . . .

Overall Score   %   _____

Remnant Risk   %   _____

(Section [B] score – Section [C] score)

6.3 INITIAL EVALUATION—ESTABLISHING THE POTENTIAL FOR SURFACE WATER POLLUTION

For many sites, it will be enough to conduct a simple desk study, supplemented by a site visit, and by a few inquiries to past landowners and to the National Rivers Authority (Section 6.2[A]).

This desk study will obviously focus on:

● the site's proximity to surface watercourses, and
● whether or not the local topographic and other conditions will permit site run-off to enter streams, rivers or lakes.

A site visit, at this stage, should show if the site currently is creating visible water pollution and also indicate the quality of waters in neighbouring surface water bodies. If the statutory water control body is then approached, fuller details on local surface water conditions are easily obtained.

The sole factor left to assess is whether the site is likely to contain those more mobile and soluble contaminants which will pose the greatest risks to water qualities. This is done by predicting the likely contamination created by the site's past uses and comparing this to the contaminant groups and concentrations in List 6.1 (Section 6.2). To obtain details of unusual and site-specific contaminants (such as—say—the organotin compounds, derived from the anti-fouling paints applied to ship hulls, which occur widely in many ship repair and breaking yards) will often call for advice from former owners and/or employees and can be difficult to achieve in the brief time usually allowed for an initial evaluation.

As noted earlier (at the start of Chapter 5), the contaminant groups and trigger concentrations given in List 6.1 are not based on new research findings, but are those contaminants and concentrations which experience has shown can create water pollution effects in especially sensitive situations.

There obviously must always be difficulty in selecting any fixed concentration of a contaminant to indicate probable water pollution concerns, since a critical factor (that of the dilution with cleaner water, which will take place when the contaminant enters a body of surface water) remains unknown. If, however, a cheap and simple risk assessment approach is ever to be possible, this difficulty has to be accepted, and minimised by deliberate selection of conservative concentration values.

List 6.1 also has its trigger values expressed mainly in terms of "total" concentrations of solid contaminants. This will seem surprising, given the obviously greater importance of mobile, soluble and leachable contaminants and the advice noted in Table 3.1. However, it has to be accepted that the industry-wide emphasis given by the ICRCL choice of guidelines (ICRCL, 1987 and Table 1.3) has been such that most contaminated site investigations list relatively few water analyses, and very seldom indeed provide information on contaminant mobilities and solubilities. To permit the assessment method to be widely utilised, it is necessary to allow for the type of contaminant information which is most likely to be available.

Weightings and scoring options for the contaminant groups of concern are given at the end of List 6.1 (Section 6.2).

Whilst this initial evaluation is biased towards surface water effects, it should not be forgotten that the presence of a near-surface groundwater (should this itself be polluted) can be a commonplace cause of surface water pollution, in at least some climatic and seasonal conditions.

6.3.1 Site examples

On the basis outlined in Section 6.2[A] it is possible to consider the three example sites (Appendix I) and judge whether any has the potential to incur significant concerns of surface water pollution.

From the information available on Site 1 (the light industrial site) it is apparent that no concerns of surface water pollution occurring would be well founded. Despite mobile and water soluble contaminants being left, after the completion of the reclamation, there are no nearby surface watercourses, and the local topography would not encourage water-borne contamination to migrate off-site. Thus no need exists for any further and more detailed assessment of this particular environmental liability.

For Site 2 (the former iron foundry) the same judgement cannot be made. This area of land is immediately adjacent to a clean river, and topographic slopes (despite the final terraced landform) certainly would not prevent some run-off reaching the river in very wet periods. It is true that the upper 2.0 m of the site is known to be composed of clean gravelly sands and so run-off should be clean, but at deeper levels soil contaminants still exist, and in fills which contain groundwaters. Groundwater also exists in the underlying natural sands and gravels and may be in continuity with the fill waters. Thus the possibility of contaminants, leached from the iron foundry fills, moving, via groundwater, to the river has to be accepted. Further evaluation of this site's potential for creating unwanted surface water pollution liabilities inevitably has to be undertaken.

For Site 3 (the former oil storage depot) surface water pollution risks are quite obvious. Not only does the site border on a clean stream, but the coastal waters are only some 200 m away, and a site visit soon indicates that oily contaminants do bleed into the stream (on the downstream end of the site) and then can be seen entering the coastal waters. In this case, it is obviously necessary to undertake a complete surface water pollution risk assessment.

As these examples illustrate, the initial evaluations are merely cheap and rapid screening tests to establish if further evaluation and costs have to be invested, or whether it will be defensible to terminate a particular risk assessment.

6.4 QUANTIFYING THE PROBABLE MAGNITUDE OF ANY FUTURE SURFACE WATER POLLUTION LIABILITY

This part of the evaluation (Section 6.2[B]) calls for information on the site's condition. In particular the following have to be known:

- the occurrence of contaminants listed in List 6.1,
- the concentrations of these contaminants, and
- whether or not these contaminants are mobile, liquid or water soluble.

The last requirement is, of course, especially important. Very high concentrations of (say) "total" copper, where these are locked into silicate bonds in a metal works slags, will be far less important than will much lower concentrations of very mobile copper salts produced by the spillages of (say) timber preservatives.

As noted earlier (at the beginning of Section 6.3), this level of contaminant detail is seldom available in UK site investigation records, where "total" concentrations are usually all that are listed. Thus it usually will be necessary either to conduct leachability tests on the solid contaminants (Lewin and Bradshaw, 1993) or, more commonly, to use the quality of local near-surface groundwaters as an indicator of contaminant leachability. Groundwater which is of local origin (and not flowing through a site and so representing off-site chemical quality conditions), and has been in contact with the solid contaminants for a meaningful time, will tend to dissolve more readily soluble contaminants, and so indicate the order of leachability effects which can be anticipated. Care, however, has to be taken with this approach, as it is quite possible to be misled by ponded and recent rainfall infiltration, which often collects, as essentially clean water, below concrete surface slabs and above near-surface obstructions. This intercepted infiltration is not necessarily a useful indicator of the leachability of solid contaminants.

The assessment method detailed in Section 6.2[B] allows the impacts both of suspended contaminants (carried off a site's surface in storm run-off conditions) and of dissolved contaminants (transported off-site via groundwaters) to be quantified.

An important aspect of the evaluation is to judge whether or not the site investigation has been adequate to provide the information needed to form valid judgements. In cases where this has not been achieved, defining the minimum additional information to fill the information gaps is necessary. Clients will usually agree to additional work in cases where the importance of the extra detail can be demonstrated.

6.4.1 Site examples

Site No. 2

Initial evaluation of the iron foundry site (Appendix I) had been that the site certainly did have the potential to create surface water deterioration, though the placing of the provably clean capping of gravelly sands obviously had removed any concerns that suspended contaminants could be carried by storm run-off into the adjacent river.

When the part [B] evaluation was completed, it became obvious that these initial concerns were less well based than had been thought. The iron foundry fills (Table AI/3) (which still exist from depths of 2.0 m to 4.0 m in the site) had been well investigated and were neither especially acidic or alkaline, and had near-neutral pHs. No free oils, mobile tars or solvents were proved to exist, nor had any unusual contaminants been produced by the iron-making processes. The usual range of metallic and sulphur contaminants (Groups B and C) did, of course, occur pervasively in the lower 2.0 m thickness of these fills, but the measured concentrations of these were never higher than the values noted in List 6.1. Thus if the solubility of the contaminants in the foundry fills had been established, it would have been possible to allocate a very low probable magnitude (-10) for any future water pollution.

However, the lack of quality detail on the groundwaters which occur in the foundry fills, the uncertainty over whether these groundwaters were directly in contact with groundwater in the underlying sands and gravels, and a complete lack of knowledge on the local groundwater flow directions (to the adjacent river?) made it clear that the site investigation had been inadequate. Thus the conservative judgement was that the probable magnitude would be between -20 and -55. The higher score was achieved by ascribing worst-case values to questions 2 and 4 (Section 6.2[B]).

Since experience indicates that such foundry fills tend only to be very slightly water soluble, it could be argued that this judgement is over-severe. However, ascribing a penalising score does emphasise that information deficiencies exist, and that it is in the client's best financial interest to have such doubts resolved by a minor additional site investigation. In this case three shallow boreholes (each 6 m deep), together with chemical analyses of water samples, both from the foundry fills and the natural sands and gravels, is all that is required.

Site No. 3

For the former oil storage site (Appendix I) the initial evaluation had been that the potential for surface water pollution was high. Quantifying the magnitude of the liability confirmed this. Free oils, gas works wastes, high soil sulphate levels, and locally very acidic ground conditions did occur (Tables AI/5 and AI/6) together with a polluted groundwater in the sand surface capping. On the basis detailed in the scoring values of List 6.1, a high liability magnitude (of -70) appeared to be well founded. Concerns over information deficiencies were not relevant in this case, since site investigation had been very complete and well conducted.

6.5 RISK REDUCTION ACHIEVED BY REMEDIATION

Site reclamation is not always directed to remove the liability concerns of greater importance. Ideally this, of course, should never be so, and developers should, in principle, always select the most appropriate reclamation approach on the basis of the site investigation details and an evaluation of the liability risks which could arise (Sections 6.2[A] and [B]).

However, it is surprising how often this is not done, and how frequently one particular contaminant liability concern is emphasised to the detriment of other equally or more important concerns.

Thus the initial judgement in Section 6.2[C] is to establish whether or not remediation has been properly targeted. If this proves not to have been the case, then it is likely that remediation benefits will not be enough to remove concerns of a particular environmental liability.

6.5.1 Site examples

This situation was indeed the case for the former iron foundry site (Appendix I). No attention had been paid to groundwater qualities or groundwater flow directions, and so it could not be established (in the risk assessment) whether or not the site would be free from the risks of creating surface water pollution. Continuing with the final stage [C] of the risk assessments process in such conditions is unproductive, and all that can be done is to advise the client that an unquantifiable risk situation exists and that additional sub-surface information is essential.

For the former oil storage depot site (Appendix I), no such uncertainty exists. Surface water pollution concerns were obviously of prime importance to the developer's advisers, and the remediation solutions adopted were not only adequate to resolve the existing water contaminant problems, but were controlled and monitored to produce very convincing proof of the quality of achievement.

Areas of polluted groundwater were defined, and then isolated by cut-off walls before groundwater removal and treatment commenced. These more contaminated areas were then excavated down to (and a short distance into) the sandstone rock, so that soil washing could remove more viscous oils trapped in soil pores and around sand grains. Testing of the washed sands was then carried out, to confirm that future leaching by rainfall infiltration would not recreate the oily pollution of local groundwaters. Polluted groundwater in the vicinity of the former gas works proved more difficult to treat successfully, and so some 15 000 cubic metres of this water had to be removed for off-site chemical treatment.

Quality control records to prove the reclamation achievement were retained, and proved to be convincing. Finally, a post-reclamation survey of groundwater qualities successfully indicated that this liability concern had been fully resolved. Completing the final (Section 6.2[C]) part of the risk assessment for this site thus gave rise to a very high evaluation score (+ 100). Remnant risks of surface water liability occurring thus can be discounted as the Part [C] positive score is larger than the Part [B] negative score.

6.6 CONCLUSIONS

It is not uncommon for developers to be prosecuted for polluting surface waters. The guilty parties suffer not only the substantial fines, which courts choose to hand down, but also are harmed by adverse press comment and by damage to their professional standings.

Yet, almost every such prosecution has arisen from instances of careless site control. Stockpiles of excavated soils are often left so unprotected that polluted run-off develops in wetter weather, and lagoons (to hold oils and other fluids) are sometimes so poorly constructed that leakages develop. These types of failures are criminal offences, yet cleaning up their consequences is usually not a difficult task.

Matters, however, would be much less simple if contaminated land had been the cause of surface water pollution. Proving which particular area, or level, in a site caused the problem can be difficult, and rectifying the consequences will obviously be more problematic when pollution reoccurs every time it rains. Remediation costs then tend to be very sizeable, and reputable developers will obviously wish not to be exposed to such consequences.

The three example sites form a reasonable spread of the contaminated land conditions which occur in the UK.

7

Groundwater Pollution Risk Assessment

7.1 GROUNDWATER POLLUTION

Polluted shallow groundwater is commonplace on former industrial sites, especially where variable fills and wastes now surface the land.

This made ground is usually typified by considerable differences in soil compaction. Some layers are loose and very porous, whilst others are firmer and more compacted. As a consequence, rainfall infiltration can often be impeded, and perched waters can occur over clayier and less open layers, or above man-made obstructions, such as buried floor slabs. With time, this impeded water can become quite polluted, usually with more easily leached oil residues and with organics from degrading wastes.

Few reclamation problems usually arise from this type of near-surface polluted water. All that is normally required is to trench the site, lead the shallow water to a sump, and then either treat the drainage or obtain consent to discharge it to a sewer. Little or no deeper groundwater concern will usually occur, though it is obvious (from the amount of oily and contaminant stainings often visible on top of clay bands) that many contaminated sites certainly have the potential to cause significant groundwater pollution. All that has prevented this occurring more commonly are the limitations on downward fluid movements in variably filled ground.

However, where deliberate or long-term tipping or leakage of liquid wastes has taken place, deeper contaminant movement is much more likely and, in some more vulnerable situations, this can reach and pollute deeper groundwaters. The recent (1993) and well-publicised legal action by Cambridge Water Company against Eastern Counties Leather is a topical example of the problems that can arise.

Groundwater pollution is, of course, a criminal offence, in exactly the same way as noted for surface water pollution (Chapter 6, Section 6.1). The same punitive fines and penal sentences exist, if a landowner or occupier is proved to be guilty. Civil actions could also result, if pollution of groundwater adversely affected local residents and it could be thought that this would make groundwater pollution a concern equal to that of polluting surface waters.

In fact, groundwater pollution should be seen as a distinctly more serious problem, since the financial consequences are invariably much greater. Surface water can always be seen; taking samples from whatever locations are necessary to identify pollutant inflows is not a difficulty; and remediation (whilst certainly expensive) calls only for widely available plant and equipment and usually can be completed in relatively short periods. In contrast, groundwater resources are never directly visible; their locations and qualities are only known in the relatively few observation boreholes which budgets have allowed; water flow paths and flow rates are unlikely to be uniform and may be greatly influenced by unpredicted variations in deeper and non-visible strata; and remediation is likely to call for specialist contractors and equipment and could last for months or even years (Childs, 1985). Thus remediation costs are likely to be far higher than those for surface water quality restoration. These cost burdens should discourage most developers from acquiring any land which could present more severe groundwater pollution problems.

7.2 GROUNDWATER PROTECTION POLICY

The practical difficulties of dealing with water resources which are not directly visible have forced the National Rivers Authority (NRA) to adopt a protection policy based on a simple and easily understood risk assessment approach (1992). Since this will form the basis of NRA reaction to any planning applications involving reclamation of contaminated sites, it is worthwhile considering the policy in some detail.

The National Rivers Authority has rightly taken the stance that contaminated sites very often have the potential to create groundwater pollution, and that—in some cases—this potential hazard could prove to be an unacceptable risk.

The reclamation of example Site 3 (Appendix I) for example, would almost certainly be seen as likely to create groundwater pollution, particularly if—as site investigation data suggest—the pools of oily residues, lying on top of the massive sandstone, are only prevented from moving deeper into the rock by the heavier oil smearings in its top few millimetres. Since reclamation proposals do entail opening the site and removing all oily contaminated soil and rock, the statutory control body would require convincing proof that this would not allow oily fluids easier passage into freshly exposed rock.

To move from the relative uncertainty of believing that a hazard could exist, to the more certain identification of situations which could pose various degrees of groundwater pollution risks, necessarily forces note to be taken of the interactions of three controlling factors:

- the types of strata (solid rock or superficial deposits) which underlie a site, and whether these are water storage and transmission horizons,
- the nature of the soils which overlie the deeper strata, and
- the thickness and composition of unsaturated materials which occur above any groundwater level.

Three types of strata adequately cover the natural variations which occur, i.e.:

Type A   Very permeable and able to yield enough water for large-scale public and industrial supplies. Typically higher water velocities occur, and so any pollution could spread widely from a source. The Chalk (southeast England) and the Magnesian Limestone (Durham and Yorkshire) aquifers would be typical Type A horizons.

Type B   Less permeable and productive aquifers, important mainly for local supplies and as the source of base flows to rivers in periods of reduced rainfall. The sands and gravels which underlie example Site No. 2 (Appendix I) would fit the Type B criteria.

Type C   Non-aquifers with very low permeability and containing only small water volumes. Usually these will be clays, shales (such as those below example Site No. 1) or especially massive rock types.

Identifying the strata as one of these three types allows the NRA to concentrate on protecting more valuable groundwater resources. However, it would not be enough to identify that a planned contaminated land reclamation lies above a Type A or a Type B aquifer and then assume that a meaningfully high groundwater pollution risk would result. Note has to be taken of other factors which will influence the likely outcome.

A particularly important variable would be the nature of any soils which overlie the aquifer, since these often will provide a protective barrier between surface reclamation and the deeper water. Three soil classes have thus been defined to allow for the range of effects which can influence the physical flows of aqueous and liquid contaminants and chemical changes which can occur during flow.

Some soils (Class 1) have little or no protective capacity, because they are either very thin or are particularly permeable to liquid flows. The usual lack of sizeable clay or silt fractions, in these soils, precludes any meaningful chemical or physical improvement of contaminated fluids.

Much greater protection is gained if thick and clay-rich soils (Class 3) exist above an aquifer. These can either divert contaminating infiltration into horizontal outflows (Cairney, 1993a) or can attenuate the fluid contaminants by a variety of chemical, physical or biochemical quality improvement processes. A fortunate accident of UK geology (Lord, 1991) is the widespread occurrence of thick glacial boulder clays, which have largely limited the groundwater pollution which otherwise would have been extremely severe from many of the older industrial centres.

An intermediate soil class (Class 2) with moderate permeability is also recognised.

As with all arbitrary systems, there will—of course—be soils whose classifications are debatable (e.g. clays with permeable sandy lenses).

The nature and thickness of the unsaturated materials above an aquifer's water level are also important. If these are thicker and more clay rich, the speed of contaminant infiltration is reduced, and time is available for the quality of any infiltrating fluid contaminants to be improved by chemical or physical reactions.

By combining these three factors—strata type, soil class and the nature of any unsaturated zone—the National Rivers Authority has devised as useful a method of assessing situations of "low", "medium" and "high" groundwater pollution risk as is practically possible. Figure 7.1 illustrates the NRA approach with reference to the three example sites detailed in Appendix I. This same system permits a developer to identify, in advance, whether adverse reactions (and increased remediation costs) are likely to arise, when planning permission is sought for a contaminated land reclamation.

A final NRA criteria is to take note of any groundwater abstractions which may occur in the vicinity of a proposed reclamation, and so be susceptible to adverse effect. As the control body the NRA has, of course, a statutory duty to protect the quality of any such abstraction, and has established three source protection zones. These are defined in terms of groundwater flow times to allow for the degradation of more easily decomposed contaminants to acceptable levels, i.e.:

Inner source protection zone. This is taken as the distance, from the groundwater abstraction, which will be covered in a 50-day water travel time. Usually 50 days will be enough for the more easily degraded contaminants to decay and lose their polluting impact.

Outer source protection zone. This larger area (defined by a 400-day water travel time) is specified to allow enough time for concentrations of more slowly degrading contaminants to decline. Since more rapid groundwater flow paths could exist, via a few particularly open fissures, NRA believes that this outer zone will never be smaller than 25% of the aquifer catchment which supplies water to the abstraction borehole.

Source catchment. This is the complete area of the aquifer which supplies water to the specified groundwater abstraction point. Usually this will be calculated on the basis of the long-term annual groundwater recharge, from infiltrating rainfall, needed to supply the licensed groundwater abstraction. Such calculations will—of course—only ever be best estimates.

Since NRA policy is clearly to minimise groundwater pollution in any of these three protection zones, and they will—indeed—seek to prosecute wherever evidence of groundwater pollution, *or the threat* of this, can be demonstrated, developers should be especially cautious before acquiring any site which overlies a productive aquifer and is within a kilometre or so from an authorised groundwater abstraction.

Whilst this policy has the benefits of clarity and of removing previous uncertainties, judgements based on it are *only* risk assessment probabilities and

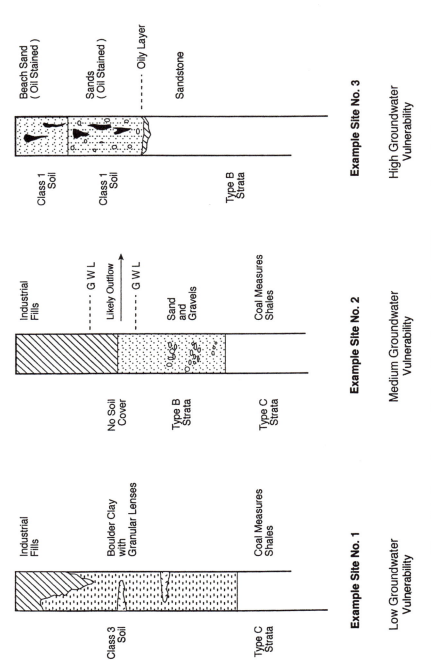

Figure 7.1. Groundwater pollution risks of example sites (GWL = groundwater level)

not unquestionable facts. Situations may arise in which the assessment policy suggested a higher groundwater pollution risk than ultimately proved to be correct, and so it certainly will not be invariably necessary for a developer to pass up a reclamation opportunity because of groundwater pollution concerns. However, proceeding in such circumstances will only be sensible if the developer accepts that he or she will inevitably have to expend extra large investigation costs, and then have to enter into time-consuming negotiations and discussions, whose final outcome might be negative.

7.3 GROUNDWATER POLLUTION RISK ASSESSMENT

Main Concern

The pollution of groundwaters, which would attract the attention of the National Rivers Authority (or equivalent control bodies).

This concern will exist if site conditions permit easy moisture infiltration into the site, and if the site is within an area vulnerable to groundwater pollution.

The concern will be greater if the site contains a particular suite of contaminants (List 7.1) and if these exist in mobile and/or water soluble forms.

[A] Potential for future liability

Tick appropriate box

1. Are the strata below the site of significance to groundwater resources?

   (a) *Yes. Site is over a Type A aquifer.* ☐

   (b) *Yes. Site is over a Type B aquifer.* ☐

   (c) *No. Site is over Type C strata.* ☐

2. If the site does lie over either a Type A or Type B aquifer, would this be at risk from any possible pollution?

   (a) *Aquifer lacks any soil cover.* ☐

   (b) *Yes. Only covered by Class 1 soils and with only a thin (<3 m) unsaturated zone above the likely groundwater table.* ☐

   (c) *Yes. But Class 2 soils do cover the aquifer, and a reasonably thick unsaturated zone (>3 m) is likely to be present.* ☐

   (d) *No. Soil cover is of Class 3 and likely to be relatively thick (>5 m). Unsaturated zone is also likely to be quite thick (>3 m).* ☐

3. Would the site topography be likely to encourage rainfall infiltration (and so the entry of any transported contamination)?

   (a) *Yes. Topography very flat.* ☐

   (b) *To an extent (e.g. surface run-off is on to lower areas from which infiltration is likely).* ☐

   (c) *No. Site topography is steep and much more likely to create surface run-off.* ☐

4. Would the materials surfacing (and within) the site be likely to encourage rainfall infiltration?

    (a) *Yes. Loose, granular and probably very permeable materials.* ☐

    (b) *To an extent (e.g. granular materials which are well compacted).* ☐

    (c) *No. Site surfaced by cohesive and/or impermeable materials.* ☐

5. Does the site have the potential to pollute groundwaters because List 7.1 materials are likely to be present in high concentrations?

    (a) *Yes. Site past use makes presence of List 7.1 materials very probable.* ☐

    (b) *To an extent. Site likely to contain List 7.1 materials, but probably not in high concentrations.* ☐

    (c) *No. Site past use makes presence of List 7.1 materials in high concentrations improbable.* ☐

6. Is there any evidence, or suspicion, that the site already is a locus of groundwater pollution?

    (a) *Yes. Specify details below.* ☐

    (b) *No evidence available.* ☐

    (c) *Evidence suggests site is not currently polluting groundwaters. Specify details below.* ☐

Q6. Evidence of effect site currently has on groundwater.

. . . .
. . . .
. . . .

---

7. Is the local groundwater known to be of high quality currently? (NRA and research body sources to be contacted)

    (a) *Yes. Specify details below.* ☐

    (b) *Information not available.* ☐

    (c) *No. Specify details below.* ☐

Q7. Details of groundwater current quality.

. . . .
. . . .
. . . .

8. Does groundwater quality monitoring take place near the site?

(a) *Yes. Specify location and details below.* ☐

(b) *Information not available.* ☐

(c) *No. Monitoring known not to be taking place.* ☐

Q8. Details of any groundwater quality monitoring.

. . . .

. . . .

. . . .

If the answers to all the above questions are "No", terminate groundwater pollution liability assessment at this point, and accept that the liability risk is likely to be negligible.

For other results, summarise the overall evaluation below. Justify clearly any decision to terminate groundwater pollution liability assessment.

. . . .

. . . .

. . . .

. . . .

LIST 7.1

Substances (expressed mainly as solid concentrations) which may give rise to groundwater pollution

Group A

Very acidic compounds   (pH < 5)
Very alkaline compounds   (pH > 9)

Note: pH changes which may be predictable (e.g. via oxidation if site disturbance occurs).

Group B

Arsenic salts	>	40 mg/kg
Boron (water soluble)	>	3 mg/kg
Cadmium salts	>	15 mg/kg
Chromium salts (as Cr VI)	>	30 mg/kg
Copper salts	>	300 mg/kg
Lead salts	>	1500 mg/kg
Mercury salts	>	20 mg/kg
Selenium salts	>	6 mg/kg
Zinc salts	>	500 mg/kg

Note: the very variable solubilities of different compounds of the same metals. High solid concentrations need *not* indicate any water pollution risk. Direct or indirect solubility proof is invariably essential for risk scoring (Q B2).

Group C

Sulphates	> 10 000 mg/kg
Sulphides	> 1 000 mg/kg

Note: the very variable solubilities of different sulphates and sulphides and the need for solubility proof in risk scoring.

Group D

Free oils or oil residues	if visibly present (> 1000 mg/kg)
Landfill leachates	if visibly present
Mobile tars/tarry liquors	if visibly present
Organic solvents	if noticeably present
Cyanides	> 500 mg/kg
Phenols	> 200 mg/kg
PAHs	> 5000 mg/kg

Note: the mobility of this group poses a more definite water pollution risk than is the case with other groups. Thus greater weight must be given to the occurrence of excessive concentrations of substances in this group.

Group E

Other site-specific compounds or wastes which are known to be water soluble:

e.g. DDT
    Pesticides and residues   > 20 mg/kg

| Phosphates | >20 mg/kg |
| Timber preservative fluids etc. | >20 mg/kg |

Note: local site history/use should be employed to indicate which site-specific compounds are of interest.

Scoring Options	*SCORE*
1. Group D substances and those of other groups present in excessive concentrations and throughout much of the site. All, or most, contaminants prove to be leachable/mobile to a significant extent.	50
2. "Hot spots" of at least three groups of contaminants are present at excessive concentrations. Leachability/mobility is only partial.	30
3. "Hot spots" of at least three groups of contaminants are present. Leachability/mobility is relatively slight, and has not led to pollution of near-surface groundwater.	20
4. Some individual contaminants occur at highish concentrations. Leachability/mobility proves to be nearly negligible.	10
5. Some individual contaminants do occur, but either at concentrations below those cited on List 7.1 and/or leachability/mobility proves to be essentially negligible.	0

Note: if contaminated layer proves significantly thick (>2.5 m) increase scores of options 2, 3, 4 and 5 by 10 points.

If contaminated layer proves to be relatively thin (<300 mm) reduce the scores of options 1, 2, 3 and 4 by 10 points.

[B] Probable magnitude of a future liability

Score (as negative values)

1. Do List 7.1 substances exist in the site, and at high concentrations?	(a) Yes.	10
	(b) To a limited extent.	5
	(c) No.	0
2. If List 7.1 substances exist, at high concentrations, will these be in mobile/water soluble forms? (see end of List 7.1 for scoring options and note thickness of any contaminated layer)	(a) Yes. Specify evidence below.	50–30
	(b) To a limited extent. Specify evidence below.	30–20
	(c) No. Specify evidence below.	10–0

*Score (as
negative
values)*

Q2. Mobility/solubility evidence.

. . . .

. . . .

. . . .

3. Would local conditions permit high rates of infiltration into, and through, the site to enter an underlying aquifer? (include data noted in Q A3, A4)

   (a) *Yes.*   10

   (b) *To a limited extent.*   5

   (c) *No.*   0

Note: distinguish whether pre- or post-reclamation conditions are being judged.

4. Would local circumstances make any such infiltration significant, in terms of groundwater quality protection? (include data noted in Q A1, A2, A7)

   (a) *Yes.*   10

   (b) *To a limited extent.*   5

   (c) *No.*   0

5. Is there evidence that groundwater pollution has already occurred on site?

   (a) *Yes. Specify evidence below (e.g. local perched groundwater quality, leachability data).*   10

   (b) *Information inconclusive. Specify details below.*   5

   (c) *No. Evidence is that no such pollution exists. Specify details below (e.g. local perched groundwater very clean).*   0

Q5. Evidence of groundwater pollution.

. . . .

. . . .

. . . .

Score (as
negative
values)

6. Is the Site Investigation data adequate for the above questions to be answered with confidence?
    (a)   Yes.   0

    (b)   *No. Specify particular data deficiencies below.*   10

Q6. Information deficiencies.

. . . .

. . . .

. . . .

Advise Environment Division Director if risk evaluation has been biased because of data inadequacies.

<u>Overall Score</u>   %   _____

[C] Risk reduction achieved by site remediation

Score (as
positive
values)

1. Have the remediation works properly focused on preventing/ reducing groundwater pollution risks?
    (a)   Yes.   10

    (b)   *Not entirely. Specify data below.*   5

    (c)   *No. Specify data below.*   0

Note: if probable sources of groundwater pollution (if this is feasible) have not been properly targeted, then remediation benefits are unlikely to be adequate.

Q1. Remediation focus on groundwater pollution risks.

. . . .

. . . .

. . . .

Score only *one* of the remediation solutions—Q2 to Q5.

*Score (as
positive
values)*

2.(a) Have *all* the mobile/soluble List   (a) Yes.   70
      7.1 substances been removed
      from the site (and particularly   (b) Not entirely. Specify data below.   35
      from any fill mantle that caps the (c) No. Specify data below.   0
      site) or reduced to below the
      critical concentrations? (see
      List 7.1)

Q2(a). Mobility/solubility evidence.

. . . .
. . . .
. . . .

   (b) If all mobile/soluble substances   (a) Yes. Specify evidence below.
       appear to have been removed (or   20
       reduced to below the critical   (b) Not conclusively. Specify evidence
       concentrations) does quality   below.   10
       control or post-remediation
       monitoring proof of this exist?   (c) No. Specify evidence deficiencies
                                              below.   0

Q2(b). Quality control/monitoring proof.

. . . .
. . . .
. . . .

3.(a) If the mobile/soluble List 7.1   (a) Yes. Specify details below.   50
      substances have been left on the
      site (particularly in any fill   (b) Not entirely. Specify details below.   25
      mantle that caps the site) do
      works exist to encapsulate and/   (c) No. Remediation was not
      or immobilise these?   adequately designed to
                                             encapsulate/immobilise the
                                             substances. Specify deficiencies
                                             below.   0

Q3(a). Adequacy of encapsulation/immobilisation.

. . . .
. . . .
. . . .

(b) If encapsulation/immobilisation has been carried out, does quality control and/or post-remediation monitoring proof of effectiveness exist?

(a) *Yes. Specify details below.*   20

(b) *Not conclusive. Specify details below.*   10

(c) *No. Specify deficiencies below.*   0

Q3(b). Quality control/monitoring proof.

. . . .
. . . .
. . . .

4.(a) If the List 7.1 substances of concern have neither been removed from site, nor encapsulated/immobilised, have "passive" protection works (e.g. markedly reducing the permeability of the site surface by installing concrete or similar cover) been included as a solution?

(a) *Yes. Specify details below.*   25

(b) *To an incomplete extent. Specify details below.*   15

(c) *No. No such protection included.*   0

Q4(a). Details of "passive" protection works.

. . . .
. . . .
. . . .

(b) If any such "passive" protection has been included as the main risk reduction factor, does it appear likely to be effective? (note probable durability)

(a) *Yes. Detail reasons below.*   20

(b) *Only to a partial extent. Detail reasons below.*   10

(c) *No. Detail reasons below.*   0

Q4(b). Likely effectiveness of "passive" works.

. . . .
. . . .
. . . .

5.(a) If the remediation consisted of a particular mix of removal/ encapsulation and immobilisation/plus "passive" protection, does this appear likely to be effective in the long term?

   (a) *Yes. Details and reasons below.*   **50**

   (b) *Not entirely. Details and reasons below.*   **25**

   (c) *No. Details and reasons below.*   **0**

Q5(a). Mix of remediation techniques and likely effectiveness.

. . . .

. . . .

. . . .

(b) In any such case, is there quality control and/or post-remediation monitoring to prove the effectiveness of the remediation?

   (a) *Yes. Specify details below.*   **20**

   (b) *Evidence not conclusive. Specify details below.*   **10**

   (c) *No. Specify details below.*   **0**

Q5(b). Quality control/monitoring proof.

. . . .

. . . .

. . . .

Overall Score   %   _____

Remnant Risk   %   _____

(Section [B] score – Section [C] score)

7.4 INITIAL EVALUATION—ESTABLISHING THE POTENTIAL FOR GROUNDWATER POLLUTION

The same type of simple desk study (Section 6.2[A]), as that outlined for establishing the potential of surface water pollution, is adequate to identify those contaminated sites which are likely to pose greater risks to groundwater resources.

Usually all that is necessary is to identify from Geological Survey and Ordnance Survey maps (and from a brief site visit) whether a productive aquifer underlies the particular site; whether infiltration of rainfall can easily enter the site's surface; and if the site is likely to contain those more mobile and soluble contaminants, which either could be carried down by the infiltrating rain, or might have been spilled or tipped as liquids in such quantities that groundwater pollution would be likely.

Responding to the final questions (does any suspicion of groundwater pollution exist, can monitoring records confirm this suspicion, and does groundwater monitoring occur in the vicinity of a particular site) can be more difficult. Groundwater monitoring is expensive to set up, and relatively few monitoring boreholes exist. Thus it usually is worthwhile to check not only with the statutory water control body, but also with any local universities which may be carrying out groundwater research. However, in many cases, particularly over the less important Type B aquifers, it will not be possible to obtain all the information that is needed, and so this initial assessment will often be less precise than was that for surface water pollution potential risks.

7.4.1 Site examples

When the three example sites (Appendix I) are considered, it is quite obvious that the former light industrial site (No. 1) poses negligible groundwater pollution risks. The site is underlain by a good thickness of Class 3 clay soils, below which are Type C strata (shales of the Coal Measures formation). Only small collections of perched groundwater occurred on the glacial clays and the Coal Measures Shales proved to be entirely dry. Thus it is defensible to terminate the groundwater pollution risk assessment at the end of this initial stage.

The same decision is possible for the former iron foundry (Site No. 2). There is indeed a Type B minor aquifer (the sands and gravels) near the site's surface, but this is again underlain by dry and thick Type C shaly strata, which will almost certainly preclude deeper migration of polluted waters. The need to establish whether waters in the sands and gravels are polluted or clean, and— if, as appears likely—these waters are in continuity with the adjacent river, has already been decided (Chapter 6, Section 6.4.1).

Thus there is no necessity to continue with the groundwater pollution risk assessment. When the already defined additional site investigation (i.e. three shallow boreholes) results are available, this decision can of course be reviewed.

The oil storage site (No. 3) is obviously more likely to be a cause of groundwater pollution concerns. Only Class 1 soils lie above the sandstone rock, which itself appears to be a Type B aquifer, likely to discharge into protected coastal waters. Thus a good deal of further evaluation is needed in this case.

7.5 QUANTIFYING THE PROBABLE MAGNITUDE OF ANY GROUNDWATER POLLUTION LIABILITY

To complete part [B] of the assessment (Section 7.3) calls for site investigation details, and the approach adopted follows that already outlined in Chapter 6 (Section 6.4). The contaminant suites that would be of concern (List 7.1) are indeed identical to those used in the surface water risk assessment, since the same importance necessarily has to be ascribed to particularly mobile or soluble contaminants.

On the basis outlined, the oil storage site (Site No. 3) is certainly of concern. List 7.1 contaminants do occur in high concentrations, and mobile oils are obviously present over large areas of the site (Tables AI/5 and AI/6). Rainfall infiltration is obviously easy, given the level site and the capping of uniform and open imported beach sands. Additionally the effects of groundwater pollution could be important on the qualities of the adjacent coastal waters. The site investigation, which had been very comprehensive, did, however, indicate that such pollution had not, as yet, affected waters in the sandstone aquifer. The fortuitous clogging of the pores and cracks in the uppermost surface of the sandstone, by heavier oil residues, had, so far, prevented recent downward movement of oily and other mobile contaminants. However, the planned excavation of contaminated sands and rock could remove or reduce this rather uncertain protection. Thus a relatively high negative score (-75) seems appropriate. This reflects the current lack of deeper groundwater pollution and the fact that no groundwater abstractions exist in the area. Thus the effects of any pollution which did occur would not be as serious as could be the case, though polluting protected coastal waters cannot be viewed as a matter of no consequence.

A cautious private-sector developer would be most unlikely to acquire such a site, unless the projected profit margins were so high that a careful, thorough and closely monitored reclamation (whose costs would be far higher than normal) could be afforded.

7.6 RISK REDUCTION ACHIEVED BY REMEDIATION

Section 7.3 part [C], outlines how remediation benefits can be quantified. A critical first judgement is to decide whether or not remediation has in fact been targeted to remove the particular concerns of groundwater pollution taking place. If this has not been the case, it is unlikely that the reclamation will have adequately reduced liability concerns, and it will usually be unproductive to continue with this part of the assessment.

In the particular case of the oil storage site, water pollution was undoubtedly recognised as a matter of major significance, and the reclamation aim was to entirely remove water pollution sources (Appendix I). Since the remediation work was closely monitored and recorded, it proved possible to have a high confidence in the claimed achievements, and so ascribe a very high risk reduction score (of essentially 100).

In this particular example assessment has to be directed not only to the direct removal of contaminated soils/rock and polluted near-surface groundwaters (reclamation stages (a) to (d)—Appendix I, Site No. 3), but also to the decision to construct an above-ground encapsulation for untreatable gas works wastes and some more oil-soaked sands and rock. The details of this encapsulation are considered in more detail in Chapters 11 and 12.

Comparison of the risk reduction score, with that earlier derived for the magnitude of a likely future liability, demonstrates that this particular site has been rendered safe for its planned re-use, and that no meaningful concerns of groundwater pollution will be valid.

7.7 CONCLUSIONS

The possibility of a contaminated site creating groundwater pollution is a particularly serious concern. Efforts, thus, should always be made to establish whether or not such risks are significant before any derelict site is purchased. It also is important to keep groundwater pollution concerns in the forefront when reclamation strategies and working procedures are decided.

Use of the NRA approach for establishing groundwater vulnerability, together with the three-part assessment system detailed above, permits the relative risks posed by different sites to be quantified without particular difficulty.

8

Area-Wide Air Pollution Risks

8.1 LEGAL BACKGROUND

Concern over the health impacts of air pollution have been voiced since the early 17th century (Evelyn, 1661). Governmental reactions were first to minimise the gross impacts of the older smokestack industries (Alkali Act 1863), and then progressively to reduce more specific health concerns (Clean Air Acts of the 1950s; Health and Safety at Work Act 1974, and the Control of Substances Hazardous to Health Act 1988). This led finally to the Environmental Protection Act (1990), and the emphasis throughout has been to control emissions from still-active industry.

The Environmental Protection Act adopts the same approach by relating air discharges to various types of manufacturing industry. For the larger and potentially more polluting of these, the Act avoids focusing on any one type of pollution, and instead imposes an Integrated Pollution Control regime. In this, the effects of discharges on air, water and soil targets are all restricted and controlled. Rigorous inspection, registration and policing by HM Inspectorate of Pollution takes place, and an increasing number of legal actions over non-compliance have resulted (Anon, 1993e).

For smaller and less polluting industries, control over air pollution is left with local authorities. An evolving and less rigorous set of standards (Anon, 1993d) is applied and an important emphasis (from the viewpoint of contaminated land's likely effects) is that air pollution which is trivial or which does not affect larger adjacent populations is not seen as a priority for remediation.

Both types of industry are subjected to restrictions on the aerial discharge of prescribed substances. The current list of these (Anon, 1991a), i.e.:

- oxides and other compounds of sulphur
- oxides and other compounds of nitrogen
- metal and metalloid dusts and fumes
- asbestos and other mineral fibres

- halogens and their compounds
- phosphorus and its compounds
- particulate-rich dusts.

can be added to at the decision of the Secretary of State for the Environment. Interestingly, whilst the wording of the relevant clauses of the 1990 Act refers quite obviously to active industrial processes, the list of prescribed substances includes many contaminants which do occur in derelict sites. Thus, enforcement or legal action, on air pollution grounds, cannot be ruled out, simply because a contaminated site is no longer in active use.

However, for an air pollution event to become a matter of official concern, it does seem necessary that discharges must be more than trivial in amount, must occur adjacent to a populated area, and must continue for more than a few hours. Since local authority environmental protection departments will deal with complaints of air quality deterioration from contaminated sites, the likely series of events can be predicted as:

- complaints from local residents,
- local authority inspection,
- discussions with developer/landowner, and
- abatement and enforcement notices issued (only if discussions fail to rectify the situation).

The opportunity for discussion and for immediate and relatively inexpensive corrective action, to halt the air pollution, make it improbable that legal enforcement or court action will occur. No UK use of legal action has yet been reported for air pollution caused by land contamination.

Thus it is improbable that area-wide pollution will be a concern as serious as that of causing water pollution. Additionally, the legal penalties so far recorded (Anon, 1993d) have been very much smaller than those in cases where surface water or groundwater pollution has been proved. Despite this, air pollution is a legal offence which any reputable developer will seek to avoid.

8.2 AIR POLLUTION RISK ASSESSMENT

Main Concern

The occurrence of air pollution to an extent that would attract the attention of one of the air quality regulating bodies.

This concern will be an abnormal occurrence and will only exist if air pollution could adversely affect adjacent built-up areas, and the health of local inhabitants.

Should a particular suite of contaminants (List 8.1) exist in high enough concentrations, the concern will inevitably be greater.

[A]   Potential for future liability

Tick appropriate box

1. Is the site located adjacent to a housing/built-up area and upwind of it? (relate to prevailing wind directions)

    (a) *Yes. Housing estate, on boundary of site, is likely to be affected if air pollution occurs.* ☐

    (b) *Yes. Built-up area on boundary of site and could be affected.* ☐

    (c) *No. Adjacent land (for a 500 m distance) is undeveloped, and no air pollution targets exist.* ☐

2. Is the site so exposed that wind effects could transmit significant air pollution risks to adjoining properties?

    (a) *Yes. Site very open and exposed to prevailing winds.* ☐

    (b) *To an extent. Some protection from wind effects does exist. Specify details below.* ☐

    (c) *No. Site is well protected from wind effects. Specify details below (tree windbreaks, built-up area, etc.).* ☐

Q2. Protection from wind effects.

. . . .

. . . .

. . . .

3. Would materials (on the site surface) be likely to become windblown, and if so would these be likely to pose inhalation/ ingestion hazards?

(a) *Yes. Materials loose, easily wind eroded, likely to contain metallic contaminants (e.g. ashy materials), asbestos or other contaminants of concern.* ☐

(b) *Yes. Materials loose and are possibly contaminated (e.g. various industrial fills on surface), or ground soaked with spilled fluids or solvents.* ☐

(c) *No. Materials neither loose nor easily wind erodible. No indications that surface material will be contaminated.* ☐

4. Is the site likely to generate gaseous vapour nuisances in significant concentrations? (confirm by reference to gas/vapour risks Chapter 9)

(a) *Yes. Landfill gases and/or the gaseous products of subterranean smouldering and/or other unusual gases likely to be present in significantly high concentrations. Specify below.* ☐

(b) *Yes. Past use of site suggests volatile liquid/solvents could exist in high enough concentrations to pose air pollution risks. Specify below.* ☐

(c) *No. Evidence for gaseous/fume nuisances (in high concentrations) does not exist.* ☐

Q4. Evidence from Chapter 9, Section 9.2.

. . . .

. . . .

. . . .

---

5. Is the site already a source of air pollution?

(a) *Yes. Dust blow/off-site migration of fumes and gases is known to currently take place. List evidence below.* ☐

(b) *There is suspicion that air pollution could be caused. List evidence below.* ☐

(c) *No. Site does not appear to be creating any air pollution risks.* ☐

Q5. Current air pollution situation.

. . . .
. . . .
. . . .

6. Is the local ambient air quality likely to be invariably good?

(a) *Yes. No significant air pollution sources exist. Rural type area.* ☐

(b) *Possibly. Whilst area is developed it is mainly non-industrial and no obvious air pollution sources exist. Note local traffic conditions.* ☐

(c) *No. Area has a reduced air quality, and a number of air polluting industrial sources.* ☐

7. Is local air pollution monitoring currently in progress, or has it been carried out in recent years?

(a) *Yes. Specify details of monitoring and any known results. Specify agency responsible for monitoring work.* ☐

(b) *No. Checks reveal no such monitoring is in progress or has taken place.* ☐

Q7. Air monitoring results.

. . . .
. . . .
. . . .

If the answers to all the above questions are "No", terminate air pollution liability assessment at this point, and accept that liability risk is likely to be trivial.
For other results, summarise the overall evaluation below. Justify clearly any decision to terminate air pollution liability assessment.

. . . .
. . . .
. . . .
. . . .
. . . .

LIST 8.1.

Contaminants and conditions which may give rise to air pollution

Group	Contaminant	Conditions
A	Asbestos	If friable, exposed to wind action and abundant. Critical level of 0.2 fibres/ml of air.
B Contaminant- rich dusts	Arsenic	critical level $>$ 0.2 mg/m^3 air
	Boron	critical level $>$10.0 mg/m^3 air
	Cadmium	critical level $>$ 0.05 mg/m^3 air
	Chromium	critical level $>$ 0.5 mg/m^3 air
	Copper	critical level $>$ 1.0 mg/m^3 air
	Lead	critical level $>$ 0.15 mg/m^3 air
	Mercury	critical level $>$ 0.05 mg/m^3 air
	Nickel	critical level $>$ 1.0 mg/m^3 air
	Selenium	critical level $>$ 0.2 mg/m^3 air
	Zinc	critical level $>$ 1.0 mg/m^3 air
	Cyanides	critical level $>$ 5.0 mg/m^3 air
	Sulphates	critical level $>$ 7.5 g ingested
	Tars	no safe level due to cancer risk
	Phenols	critical level $>$ 5 ppm in air
	PCBs	critical level $>$ 0.5 mg/m^{3}air

Note: Dust/inhalation/ingestion risks will be significant *only* if site surface material is fine-grained/loose *and* if contaminant concentrations are well in excess of ICRCL threshold values for parks, playing fields and open spaces. Relative importance of contaminants can be taken from the Occupational Exposure Levels, cited above.
Concentrations are those on site boundary, *not* within site itself.

Group	Contaminant	Conditions
C Gaseous emissions	Carbon dioxide	critical level $>$ 0.5%
	Carbon monoxide	critical level $>$200 ppm
	Hydrogen cyanide	critical level $>$ 8 ppm
	Hydrogen sulphide	critical level $>$ 20 ppm
	Phosphine	critical level $>$ 0.3 ppm
	Sulphur dioxide	critical level $>$ 2.0 ppm

Note: These gaseous hazards will be more likely to pose air pollution risks if the combustion of such wastes as colliery spoils (subterranean combustion), cyanides and spent oxides, phosphorous wastes, coal/oil wastes, or wood preserved with chemicals occurs. As very high outflow rates will be required to create air pollution, the more normal decomposition of organic and other wastes is most unlikely to prove a risk. Category [D] gassing conditions—see Section 9.2, List 9.4—thus have to occur before a significant risk is reasonably identified.
The critical emmission concentrations will be those on the site boundary, *not* the concentrations in sub-surface measurement points.

Scoring Options	Score
1. If any of the three groups of contaminants is present in significant concentrations/outflow rates, *and* if the physical conditions for easy migration exist.	50
2. As [1], but with less suitable physical conditions for easy migration. Note physical nature of site surface, the existence of sealing layers, etc.	30
3. If any of the three groups of contaminants is present, but in less significant concentrations/outflow rates. Physical conditions for easy migration still exist.	20
4. If any of the three groups of contaminants is present, *but* concentrations/flow rates are less significant *and* the physical conditions for easy migration are less than ideal.	10
5. If concentrations/flow rates are below the critical levels.	0

[B] Probable magnitude of a future liability

Score (as negative values)

1. Do List 8.1 substances exist on/in the site and at high concentrations/ outflow rates?	(a) *Yes.*	10
	(b) *To a limited extent.*	5
	(c) *No.*	0
2. If List 8.1 substances do exist at high concentrations, will these be easily mobilised into air pollution risks? (see end of List 8.1 for scoring options and note longer term potential for risk)	(a) *Yes. Loose metallic/asbestos-rich materials on surface, and/or larger outflows of gases do exist. Specify evidence below.*	50–30
	(b) *To a limited extent. Specify evidence below.*	30–20
	(c) *No. Specify evidence below.*	10–0

Q2. Potential for mobilisation to air pollution risks.

. . . .

. . . .

. . . .

3. Would the local conditions permit significant off-site air pollution to occur? (include data noted in Q A2, A3, A4, A5)	(a) *Yes.*	10
	(b) *To a limited extent.*	5
	(c) *No.*	0

Score (as negative values)

4. Would the local conditions make any off-site air pollution a significant matter? (include data noted in Q A1, A5, A6, A7)

   (a) Yes. 10

   (b) To a limited extent. 5

   (c) No. 0

5. From any air monitoring data that is available, is there evidence of current or past air pollution events?

   (a) Yes. Specify details below. 10

   (b) Data inconclusive. Specify details below. 5

   (c) No. Data does not exist. 0

Q5. Information on current/past air pollution events.

. . . .

. . . .

. . . .

---

6. Is the Site Investigation data adequate for the above questions to be answered with confidence?

   (a) Yes. 0

   (b) No. Specify particular data deficiencies below. 10

Q6. Information deficiencies.

. . . .

. . . .

. . . .

Advise Environment Division Director if risk evaluation has been biased because of data inadequacies.

Overall Score   %   _____

[C] Risk reduction achieved by site remediation

Score (as positive values)

1. Have the remediation works properly focused on preventing/reducing air pollution risks?

   (a) Yes. 10

   (b) Not entirely. Specify data below. 5

   (c) No. Specify data below. 0

Note: If probable sources of air pollution (if this is feasible) have not been properly targeted, then remediation benefits are unlikely to be adequate.

Q1. Remediation focus on air pollution risks.

....

....

....

Score only *one* of the remediation solutions—Q2, Q3/Q4*, Q5.

*Score
(as positive
values)*

2.(a) Have all the substances which
could give rise to air pollution
been removed from the site, or
have their concentrations been
reduced to below the critical
levels? (see List 8.1)

(a) Yes.   70

(b) Not entirely. Specify data below.   35

(c) No. Specify data below.   0

Q2(a).

....

....

....

(b) If the substances which could
give rise to air pollution have been
removed (or reduced to below
critical concentrations), does
quality control and/or post-
remediation monitoring proof of
this exist?

(a) Yes. Specify evidence below.   20

(b) Not conclusively. Specify evidence
below.   10

(c) No. Specify deficiencies below.   0

Q2(b). Quality control/monitoring proof.

....

....

....

Note: the need to address *both* air pollution from solid contaminants and from gas nuisances.
If both Q3 and Q4 have to be addressed, reduce scores allowed by 50%.

Score
(as positive
values)

3.(a)* If solid substances which could give rise to air pollution have been left on site, are they adequately encapsulated/covered?

(a) *Yes. Specify details below.*   50

(b) *Not conclusively. Specify evidence below.*   10

(c) *No. Specify deficiencies below.*   0

Q3(a). Adequacy of encapsulation/covering.

. . . .
. . . .
. . . .

(b) If encapsulation/covering has been carried out, does quality control or other monitoring proof of effectiveness exist?

(a) *Yes. Specify details below.*   20

(b) *Not conclusive. Specify details below.*   10

(c) *No. Specify details below.*   0

Q3(b). Quality control/monitoring proof.

. . . .
. . . .
. . . .

4.(a)* If gaseous nuisances which were seen as a main potential cause of air pollution have not been removed, have adequate measures been taken to control these emissions?

(a) *Yes. Specify details below.*   50

(b) *To an incomplete extent. Specify details below.*   25

(c) *No. Measure seems inadequate. Specify details below.*   0

Q4(a). Reduction of gas nuisances.

. . . .
. . . .
. . . .

Score
(as positive
values)

(b) If such reduction of still-existing   *(a)* *Yes. Specify details below*   20
     gas nuisances took place, is there
     quality control and/or other   *(b)* *Only to a partial extent. Specify*
     monitoring proof of the   *details below.*   10
     effectiveness of the work?   *(c)* *No. Detail data deficiencies below.*   0

Q4(b). Quality control/monitoring proof.

....

....

....

5.(a) If the remediation consisted of a   *(a)* *Yes. Details and reasons below.*   50
      particular mix of the above
      techniques, does this appear likely *(b)* *Not entirely. Details and reasons*
      to be effective in the long term?   *below.*   25

                                           (c) *No. Details and reasons below.*   0

Q5(a). Mix of the remediation and likely effectiveness.

....

....

....

....

(b) In such a case, is there quality   *(a)* *Yes. Specify details below.*   20
     control and/or post-remediation
     monitoring to prove the   *(b)* *Evidence not conclusive. Specify*
     effectiveness of the remediation?   *details below.*   10

                                           (c) *No. Specify details below.*   0

Q5(b). Quality control/monitoring proof.

....

....

....

Overall Score   %   _____

Remnant risk   %   _____

(Section [B] score – Section [C] score)

8.3   AIR POLLUTION RISK EVALUATIONS

If a clear distinction is drawn between concerns over longer term air pollution (continually caused by a contaminated site) and briefer polluting episodes when land remediation is in progress, it is apparent that none of the three example sites (Appendix I) is likely to be subject to future air pollution liabilities.

Site No. 1 lacks asbestos and any significant emissions of gases and vapours. The metal-rich combustion ashes offered no air pollution opportunity in their pre-reclamation state (when they were capped by lush grass and small bushes) and their entire removal in the reclamation precluded any future concerns. The sole worry would have been during reclamation if any stockpiles of excavated ashes had dried out and become available for windblow. However, the developer took excavated materials off site each day and also had dust-suppressing water sprays available to minimise nuisances to local residents and so avoided the short-term problem.

Likewise, Site No. 2 lacked asbestos and its metal-rich foundry ashes were entirely capped, in the pre-reclamation period, with concrete floor slabs and roadways. After reclamation the well-laid clean cover was equally effective in preventing any air mobilisation of those metallic contaminants left at depth in the site. The presence of high carbon dioxide concentrations (up to 15%), found in the site during re-investigation, could well have been a cause of longer-term concern. However, detailed evaluation (Chapter 9, Section 9.3) indicated that this denser gas never was able to rise to the near-surface upper levels. As with Site No. 1, the sole concern was that of creating dustblow conditions during the site excavation and reclamation, but available records indicate that this was avoided by careful site management.

Site No. 3 has much more serious potential air pollution problems, largely because very high concentrations (up to 137 mg/litre) of volatile benzene were invariably encountered whenever the surfacing sands were excavated. In its pre-reclamation state, however, no benzene emissions could be measured at the land surface, and since removal of the contaminated groundwater (and the associated air flushing effect) preceded any excavation during the reclamation period this proved adequate to reduce benzene concentrations, in excavations, to less than 5 mg/litre of air (the maximum exposure level permitted by Health and Safety regulations) and prevented any off-site migration of the volatile hydrocarbon.

8.4   MORE EXTREME AREA-WIDE AIR POLLUTION

Whilst few contaminated sites have the potential to create air pollution of real concern to control bodies, some few do.

These sites, in the author's experience, fall into one or other of two types:

● Sites where shipbreaking or ship repair has dominated past land use.
● Sites where excessively high gas emissions are taking place, often because of subterranean smouldering.

Shipbreaking yards in the north-east and south-west of England inevitably had the need to dispose of vast quantities of asbestos lagging from ship boiler rooms. Where this sort of activity lasted for decades, the asbestos quantities appear to be measurable more in hundreds of tonnes than in any smaller quantities. Since the waste was seen (until the 1970s) as an inert and non-re-usable material, its fate was usually disposal in the yard's own tip, where it was mixed with other inert scrap metal, old piping and timber balks. Such tips today are usually abandoned and are poorly capped with soil, through which scraps of asbestos can be seen.

In windy periods, smaller pieces of weathered asbestos tend to blow around and off the tips, creating, no doubt, unacceptably dangerous atmospheres in surrounding neighbourhoods. Any developer purchasing such a site would inevitably also acquire area-wide air pollution liabilities, very sizeable reclamation costs and significant restrictions on the allowable reclamation method.

Similar problems also are obvious where ship repair was located, particularly if sand-blasting of ships' hulls was a large part of the activities. Such sand-blasting leaves behind fine silica sand, each grain of which is coated with metallic fragments and with the remnants of the anti-fouling paint used to protect ship hulls from the growths of marine parasites. Since these anti-fouling paints were composed of particularly toxic and biologically available compounds (e.g. organotin) they pose a particularly serious health hazard if ingested or inhaled. The waste blasting sands are, of course, loose, fine-grained and transportable in higher winds. As with the breaking yards, pervasively contaminated with asbestos lagging, the acquisition of such ship repair yards will usually mean inheriting area-wide air pollution concerns of serious magnitude.

The final category of land where widespread air pollution is likely to be a problem is where colliery spoils, cyanide or phosphorus-rich wastes or oily wastes are undergoing subterranean smouldering, usually because of surface fires that have migrated deeper into tipped areas. These sites (marked by vapour and smoke emissions and by cracked and settling ground surfaces) are especially unsuitable for any cost-effective reclamation and are prone to give rise to serious air pollution concerns.

8.5 CONCLUSIONS

Creating area-wide air pollution is usually the least severe of the environmental liabilities which contaminated land can pose. The only sites where this is not the case are areas of land which are especially ill-suited for re-use and reclamation.

Despite this, it still is worthwhile to carry out the risk assessment exercise, if for no other reason than to highlight where care should be taken during reclamation, to avoid short-term air pollution worries.

The assessment may, in some cases, indicate (e.g. Site No. 3—Appendix I) where widespread air pollution could result unless particular remediation care is taken.

9

Risks to Site Users from Gases and Vapours

9.1 THE PROBLEMS OF INTERPRETING GAS AND VAPOUR MEASUREMENTS

Gases and vapours are commonplace contaminants on many derelict sites, particularly where tipped wastes or subterranean smouldering occurs.

Unusually, a large amount of information exists on the properties and hazards of these lighter fluids (e.g. Department of the Environment, 1985a and b; HM Inspectorate of Pollution, 1989; Institute of Wastes Management, 1990). The risks to people (Table 9.1) and to plant populations (Table 9.2) are known in reasonable detail and published experiences of reducing these risks (e.g. Anon, 1991a) are usefully comprehensive.

Thus, it could be anticipated that evaluating gas and vapour risks would be relatively straightforward. However, this is *not* the case, as the very real interpretation difficulties, commonly encountered when measured gas/vapour information has to be analysed, discourage many investigators.

Measuring the concentrations, pressures and flow rates of gases and vapours is no longer difficult or expensive (Crowhurst, 1987). A wide range of field-portable gas concentration meters exists, and it is also straightforward to collect samples in the field, and take these back to the laboratory, for the intrinsically more accurate concentration measurements which gas chromatographic equipment will provide. When gas/vapour pressures and flow rates are required (as should be the norm), measurement is equally simple, though relatively few investigators have yet invested in the necessary equipment. The (unfortunate) emphasis on concentration data stems from older advisory documents which overlook the importance of flow and pressure information (Smith, 1993).

Thus, on one of the very routine investigations, to determine whether landfill gases affect a site, it is not difficult to establish the internal atmosphere of a particular monitoring borehole. On a specific day, this could be typified by normal oxygen levels (20.1%), traces of carbon dioxide (<0.5%), no methane content, and slight negative gas pressures. However, a week later, the same borehole could reveal entirely different conditions—extremely depleted oxygen

Table 9.1. Some gas and vapour risks to people (after Barry, 1990)

	Gas vapour	Effects and limits
1.	Carbon dioxide (CO_2)	Health affected at $>3\%$ concentration Toxic effects at >12 to 25% concentrations Asphyxiating effects at >12 to 25% Fatalities at $>25\%$ OEL (8 hours exposure) 0.5%
2.	Carbon monoxide (CO)	Health affected at >200 ppm Toxic effects at >0.1 to 1.0% Inflammable at $>12\%$ OEL (8 hours exposure) 50 ppm
3.	Hydrogen cyanide (HCN)	Health affected at <18 ppm Toxic effects at >100 ppm Fatal at >300 ppm Inflammable and explosive at 6% concentrations OEL (10 mins exposure) 10 ppm
4.	Hydrogen sulphide (H_2S)	Health affected at >20 ppm Toxic affects at >400 ppm Fatal at >700 ppm Explosive at 4 to 5% concentrations OEL (8 hours exposure) 10 ppm
5.	Methane (CH_4)	Inflammable at concentrations between 5 and 15% in air Explosive at same concentrations Asphyxiant Concern if concentration is in excess of 1%
6.	Phosphine (PH_3)	Toxic Explosive (spontaneously inflammable) OEL (8 hours exposure) 0.3 ppm
7.	Sulphur dioxide (SO_2)	Health effects at >6 to 12 ppm Toxic OEL (8 hours exposure) 2 ppm

OEL = Occupational Exposure Limit.

levels ($<2\%$), dangerously high carbon dioxide concentration (10 to 15%), explosive levels of methane ($>5\%$), and positive gas pressures.

What could have occurred between the two monitoring events? Which (if either) of the sets of gas atmosphere readings can be taken as typifying the site's gassing conditions?

A number of explanations can be advanced, since theoretical knowledge of gas generation and migration is quite comprehensive:

● changes could have occurred at the source producing the landfill gases,
● pathways from the gas source to the monitoring borehole could have altered,

Table 9.2. Gas and vapour risks to plant populations (after Barry 1991; Dobson and Moffat, 1993)

1.	Oxygen (O_2)	Plant root growth restricted if oxygen concentration $< 10\%$ Plant root growth ceases at 3 to 5% concentrations Oxygen restrictions less severe if they occur during dormant (winter) periods
2.	Methane (CH_4)	Cause of asphyxiation, but only because methane can displace and reduce oxygen
3.	Carbon dioxide (CO_2)	Toxic to plant roots, particularly in low oxygen environments Species response to carbon dioxide is very variable. Some show growth inhibition at 1% carbon dioxide, others require 20% carbon dioxide for same effects
4.	Carbon monoxide (CO)	Toxic and with effects similar to oxygen depletion
5.	Ethylene	Disturbs plant growth hormones at concentrations of 1 to 5 ppm. Root growth usually affected and so growth reduced
6.	Hydrogen sulphide (H_2S)	Toxic at concentrations in excess of 400 ppm
7.	Sulphur dioxide (SO_2)	Toxic

Note: Gas and vapour effects on plant life are likely to be highly species specific and to be interrelated to other factors (e.g. lack of moisture, waterlogging, etc.).

- reactions (during gas migration) could have modified the original concentrations of some gases,
- the buoyancy of the landfill gas could have been a factor, and
- dilution with atmospheric air could have taken place in the earlier monitoring event,

though it may not be easy to be sure which of these effects has been the actual cause.

Changes at gas generation sources are known to directly reflect bacterial activity. The higher the rate at which degradable materials are decomposed, the greater will be the concentrations and pressures of methane and carbon dioxide. However, bacterial activity is itself affected by environmental conditions at gas generation sources (i.e. moisture content in the degradable wastes, waste temperatures, acidity levels, amount of available foodstock, and whether external factors interfere with the expansion of bacterial populations—Rees and Grainger, 1982). So the second set of gas measurements could have arisen because of a rise in soil temperatures as especially degradable wastes were encountered by the bacteria, or because the chilling action of an earlier rainfall infiltration event had dissipated. If a gassing source is at relatively shallow depths (< 4 m) then the variations in gas measurements could be due to such effects. However, the rate and degree of change between the two sets of readings seem rather high for a gas source variation to be the prime cause.

Alterations in the flow paths, via which gases migrate to observation boreholes, could be a more feasible explanation. If the atmospheric pressure had been high at the time of the first set of gas readings, then upward movement of landfill gases could have been prevented, and the measured negative gas atmosphere could suggest this contention. If, just before the second gas monitoring event, the atmospheric pressures had abruptly declined, then a "pumping action" would have resulted, and so have led to higher landfill gas concentrations and the noted positive gas pressures. Since barometric changes do occur very rapidly and influence shallower gas migrations very quickly, this explanation would be worth investigating. Other pathway variations—e.g. if a saturated near-surface soil layer, which has been effectively gas impermeable during the first monitoring period, had partially drained and so become much more gas permeable by the time of the second monitoring—would also be feasible.

Reactions during gas migration also have effects, though these would be unlikely to be rapid enough to explain the changes in the measured data. Methane can be oxidised to carbon dioxide by near-surface bacterial species, and carbon dioxide itself is often preferentially dissolved in groundwaters. Thus distinct changes in gas concentrations with travel distance from the gas source are well reported (Ward *et al.*, 1993). However, these would not be convincing hypotheses in this particular case.

The physical buoyancy of landfill gases can also be a factor. Whilst landfill gas mixtures (Ward *et al.*, 1993) tend to have densities essentially similar to that of air itself, these do vary depending on whether or not the gases are water saturated or dry and hot. Since rapidly emitting landfill gases are distinctly warmer than ambient soil temperatures, these gases have much lower densities, are distinctly buoyant and so are able to rise rapidly through soil profiles. If this were a contributory cause of the observed facts it would be worth checking if enhanced soil temperatures had developed between the two monitoring events.

Finally, simple dilution with atmospheric air can easily affect landfill gas readings. This could result from the presence of local excavations—say for a road bed, or because other nearby site investigation trial pitting had been taking place—during the period of the first landfill gas monitoring.

Obviously, it would also be feasible that several of those separate effects had interacted to give the differences in the measured gas information.

The difficulties of interpreting gas/vapour monitoring information are thus obvious (Table 9.3) and this emphasises the need to form judgements on an adequately long set of gas/vapour data, which extends beyond simple concentration readings (List 9.3).

Gas and vapour monitoring is, of course, *only* carried out to establish whether future risks exist.

Table 9.3. Variability of gas measurements in a single borehole

Date	Volumetric percentage concentrations of:			Observations
	Oxygen	Carbon dioxide	Methane	
5 Oct 1991	20.4	0.3	nil	—
10 Nov 1991	16.1	1.7	0.3	Reading after sudden barometric fall (24 mbars)
4 Dec 1991	20.1	0.4	nil	—
3 April 1992	19.7	1.4	0.1 ⎤	Period of gradually
24 April 1992	17.4	2.7	0.6 ⎥	increasing soil
5 May 1992	10.4	8.7	2.5 ⎥	temperatures (from 8°
6 June 1992	2.3	12.7	4.8 ⎦	to 11 °C)
1 July 1992	0.1	2.3	5.7 ⎤	Near-surface soils
4 Aug 1992	0.1	1.3	4.1 ⎦	saturated by prolonged rainfall in early July
7 Sept 1992	0.1	16.2	5.5	Near-surface soils dried out
4 Oct 1992	12.4	4.9	trace	Soil temperature falling
5 Nov 1992	18.7	1.2	nil	Soil temperature at 8.1 °C
3 Dec 1992	20.4	0.2	nil	Soil surface frozen

Note: gas source 3 m thick refuse layer at depth of 2.3 m below land surface.

To establish risks calls for a good deal more than the gas/vapour concentration data, which is all too many investigators still collect. Pressure and flow rate information is essential if a site's gassing conditions (List 9.4, section 9.2) are to be established and it will often be necessary to:

● "map" out gassing zones (by employing grids of shallower gas concentration spike tests—usually to depths of > 1.5 m),
● employ soil temperatures to help locate gassing "hot spots", and
● distinguish similar gas types (e.g. those from landfills, peat layers and coal mines) by detailed chromatography analysis of trace constituents (Institute of Wastes Management, 1990).

Identifying the actual gassing condition that is occurring is crucial, since otherwise it is not possible to predict likely future conditions, or to decide whether or not remediation techniques can be expected to be successful. A range of these techniques exists (Card, 1993). Whilst some techniques (i.e. total excavation of the gas source materials) can entirely remove any future risks, others (preventing gas entry to dwellings and structures by installing external venting and barrier works) cannot be seen entirely as "fail-safe" protection. When the sole option is to gas-proof actual houses and other structures, remnant concerns must still persist, since routine maintenance and performance checking of the gas-proofing measures will not be practicable.

Until gassing conditions are established, it will never be possible to select the most appropriate remedial solution or even decide whether it is prudent to develop the particular site for a sensitive re-use.

Whilst the emphasis above has been on the difficulties presented by landfill gas information, it should not be forgotten that other gas and vapour sources can exist and create distinct liability risks. A similar variability exists in the readings obtained of these gases and vapours, and the same interpretative problem will occur.

The gases created by subterranean smouldering (List 9.2, Section 9.2) do increase abruptly in flow rates and concentrations with changes in atmospheric pressure, as additional air is permitted to enter a site's surface to accelerate combustion, and as a heating front intersects areas where a higher proportion of combustible material exists. Likewise the widespread problem of solvent vapours is more obvious if monitoring takes place in times of low atmospheric pressure, and when soil temperatures are at their highest annual point.

Coal-bearing rocks, whether mined or unmined, are known (Williams and Hitchman, 1989) to give rise to significant gassing problems. These usually are not susceptible to total remediation and the risks can only be minimised by active or passive gas abstraction approaches (Robinson and Grayson, 1990) or by gasproofing dwellings. The worst coal-bearing rock conditions are likely to be where shallow and still open mined voids exist below a site, since these (in periods of low barometric pressure) do give rise to large-scale outflows of methane and carbon dioxide, often at very high concentrations.

Gases and vapours present potentially the most serious and rapidly acting risks to users of reclaimed contaminated land. The high mobility, and the ability of the less dense fluids to penetrate into and collect in confined spaces, gives them a risk significance far beyond that of most solid contaminants.

If asbestos is ignored, no cases have yet been convincingly demonstrated in which human health and safety has been directly affected by solid contaminants. Yet deaths, injuries and financial losses from gaseous contaminants are routinely recorded. Whilst most of these instances arise from gases collecting in older poorly maintained sewers, or in unventilated trenches dug through contaminated land, or under older houses built over coal mined areas, it cannot be assumed that a similar intensity of incidents will not occur in future on reclaimed contaminated sites. Age is likely to degrade gas-proofing measures and make gas entry more likely.

Thus, it is imperative to establish gas and vapour risks fully, and to ensure that gas surveys are adequate in scope to allow the actual gassing conditions to be properly categorised.

9.2   GAS AND/OR VAPOUR RISKS TO INDIVIDUAL PROPERTIES AND/OR HOME OWNERS

Main Concerns

Four specific concerns could exist:

(a) explosion,
(b) asphyxiation,
(c) plant die-back in gardens, and
(d) cancer risks from vapours.

These concerns will be of greatest importance to individual property owners and adjacent landowners and will also be of interest to local authority planners and building control officers.

The presence of a limited range of gas and/or vapour sources (Lists 9.1 and 9.2) will make the concerns greater.

For these concerns, the worst-case assumption, that sensitive targets always exist, has to be made.

[A]   Potential for future liability

Tick appropriate box

1. Do biodegradable materials occur on the site? (categorise on List 9.1 basis)

   (a) *Yes. Tipped domestic refuse and/or "oil"-soaked ground occurs widely over site. (Note: "oil" is used to include petroleum and volatile solvents/liquids.)* ☐

   (b) *Yes. Tipped industrial wastes and/or thicker (> 1 m) bands of ash occur on the site.* ☐

   (c) *Yes. Scraps of timber waste and similar materials likely to exist on site, although probably in minor concentrations.* ☐

   (d) *No. Past history does not indicate that biodegradables should exist.* ☐

2. Do shallow mine-workings (coal or limestone) underlie the site to depths of up to 50 m?

   (a) *Yes.* ☐

   (b) *No.* ☐

3. Do thicker peat deposits (>2 m)　　(a)　Yes.　　　　　　□
   exist below the site?
   　　　　　　　　　　　　　　　　　(b)　No.　　　　　　　□

4. Are site conditions such that the　　(a)　Yes. Specify details　□
   more unusual gases (H$_2$S/CO/
   HCH/PH$_3$/SO$_2$) could occur?　　(b)　No.　　　　　　　□
   (categorise on List 9.2 basis)

Q4. Details of conditions which could give rise to relatively unusual gases.
. . . .
. . . .
. . . .

---

5. If any gas/vapour nuisances are　　(a)　Yes. Site materials granular and
   postulated (from Q 1 to 4), are the　　　　loose. Likely to be highly gas
   site conditions likely to encourage　　　　permeable, as local groundwater
   gas/vapour migration?　　　　　　　　　table is likely to be relatively deep
   　　　　　　　　　　　　　　　　　　　(>3 m).　　　　　　　　　　　□

   　　　　　　　　　　　　　　　　(b)　Yes. Site materials granular.
   　　　　　　　　　　　　　　　　　　Probably not highly gas
   　　　　　　　　　　　　　　　　　　permeable, as compaction
   　　　　　　　　　　　　　　　　　　relatively high and groundwater
   　　　　　　　　　　　　　　　　　　likely to occur near surface (<3 m
   　　　　　　　　　　　　　　　　　　depth).　　　　　　　　　　　□

   　　　　　　　　　　　　　　　　(c)　No. Site materials cohesive
   　　　　　　　　　　　　　　　　　　and well compacted. High
   　　　　　　　　　　　　　　　　　　saturation moisture content likely.
   　　　　　　　　　　　　　　　　　　Gas permeability should be
   　　　　　　　　　　　　　　　　　　low.　　　　　　　　　　　　□

6. Is there any evidence that gas　　　(a)　Yes. Gas emissions in ditches and
   vapour nuisances already exist and　　　excavations are obvious.　　　□
   adversely affect the site?
   　　　　　　　　　　　　　　　　(b)　Yes. Plant die-back apparent in
   　　　　　　　　　　　　　　　　　　some areas of the site. No other
   　　　　　　　　　　　　　　　　　　cause of plant harm is apparent.　□

   　　　　　　　　　　　　　　　　(c)　Yes. Distinct odours are obvious.
   　　　　　　　　　　　　　　　　　　Describe odours below.　　　□

   　　　　　　　　　　　　　　　　(d)　No. No evidence for gas/vapour
   　　　　　　　　　　　　　　　　　　nuisances is apparent.　　　□

Q6. Evidence of on-site gases.

....
....
....

If the answers to all the above questions are "No", and if an initial review of the Site Investigation data confirms this, terminate gas/vapour liability assessment at this point.

For other results, summarise overall evaluation below. Justify clearly any decision to terminate gas/vapour liability assessment.

....
....
....
....
....

LIST 9.1

Conditions likely to create landfill gas hazards (listed in decreasing order of importance).

1. Biodegradable wastes (domestic, wood pulp, paper or similar) underlie at least a significant area of the site and in thicknesses of > 4 m. Heavily "oil"-soaked ground, if to depths of 4 m, may pose a similar hazard if this occurs over a significant area of the site. (Note: "oil" is used to include petroleum products *and* volatile solvents/liquids.)
2. Thinner deposits (> 1 m but < 4 m) of biodegradable and industrial wastes and/or "oil"-soaked ground. (Note: "oil" is used to include petroleum products *and* volatile solvents/liquids.)
3. Layers of combustion ashes and/or some casting sands, particularly if scraps of more readily decomposable material (cloth, papers, etc.) exist in these.
4. Relatively scarce scraps of wood, hardboard, cloth, etc. (< 5% by volume of any fill layer which may exist) in inert matrix materials.

LIST 9.2

Conditions which may lead to relatively unusual gases and vapours being hazards.

	Condition	Gaseous/vapour hazard
1.	Solvent disposal via soakaways	Various solvents (cancer risks, fire risks)
2.	Subterranean combustion, particularly of colliery spoil or buried wood, paper waste, etc.	Carbon monoxide Carbon dioxide (toxins and asphyxiants)
3.	Combustion of cyanide wastes or spent oxides on/or below site surface	Hydrogen cyanide gas (toxic)
4.	Combustion of phosphorus wastes on or below site surface	Phosphine (toxic and explosive)
5.	Combustion of sulphur-rich material, on or below site surface	Sulphur dioxide (toxic)
6.	Decomposition of plasterboard or gypsum wastes in industrial tips; decomposition of oil wastes in landfills	Hydrogen sulphide (toxic, inflammable)

LIST 9.3

Adequacy of gas surveys

Category	Scope of survey(s)	Score
A	Concentrations of a limited range of gases, measured by field meter, and only on one occasion. No independent accuracy checks by gas chromatography (e.g. only CH_4 and CO_2 concentrations measured for landfill gas evaluations)	20
B	Concentrations of entire range of relevant gases measured and field meter results confirmed by gas chromatography. Concentrations measured only on one occasion	15
C	As [B], but measurements repeated on other occasions. Note made of atmospheric pressure variations and other climatic influences	10
D	As [C] plus measurements of gas pressures undertaken. Soil and/or gas temperatures may also have been measured	5
E	As [D] plus direct measurement of gas flow rates. Better surveys will have continued gas flow rate/concentration measurements for an appreciable period	0

LIST 9.4

Categories of gassing conditions (in increasing order of hazard)

Category	Conditions	Hazard score
A	Gas produced intermittently and often in response to climatic controls. Flow rates and pressures usually trivial, though concentrations can, on occasion, exceed critical limits (e.g. LEL for flammable gases, OEL for toxic/asphyxiant gases)	10
B	Gas produced continuously, though production rate varies (often in response to climatic controls). Flow rates and pressures usually trivial, though concentrations will often exceed critical limits (e.g. LEL for flammable gases, OEL for toxic/asphyxiant gases)	25
C	Gas may be produced intermittently and often in response to climatic controls. Flow rates and pressures, however, can often be high, and concentrations will often exceed critical limits (e.g. LEL for flammable gases, OEL for toxic/asphyxiant gases)	50
D	Gas produced rapidly in large volumes and under significant positive pressure. High concentrations of hazardous gases will normally be present, and usually in concentrations (which after dilution with air) will still pose serious explosion/fire/toxicity risks	65

LEL = Lower Explosive Level (for methane in air).
OEL = Occupational Exposure Level.

[B] Probable magnitude of a future liability

Tick appropriate box

1. Have surveys to identify gas/vapour hazards, been carried out?

   (a) Yes. ☐

   (b) No. ☐

If surveys have *not* been carried out and if Q A1, A2, A3 or A4 give positive responses, *terminate risk analysis at this stage* and define the minimal survey necessary for this site to be evaluated further.

Score (as negative values)

2. Were gas/vapour surveys conducted at the time of year most likely to reveal the worst conditions?

   (a) No. Survey(s) in winter period. 5

   (b) Position uncertain (e.g. surveys in late autumn or early spring). 3

   (c) Yes. Surveys in hotter weather. 0

3. Were surveys adequate in scope? (categorise on List 9.3 basis)

   (a) No. Only Category (A) surveys. 20

   (b) No. Category (B) surveys. 15

   (c) Yes. Category (C) surveys. 10

   (d) Yes. Category (D) surveys. 5

   (e) Yes. Category (E) surveys. 0

4. Were surveys adequate to prove the actual source(s) of any gas/vapours?

   (a) No. Detail reasons below. 10

   (b) Not entirely. Detail reasons below. 5

   (c) Yes. Detail reasons below. 0

Q4. Adequacy of surveys.

. . . .

. . . .

. . . .

---

5. Do the surveys adequately define which gassing condition exists? (categorise on List 9.4 basis)

   (a) Yes. Category (A) conditions exist. 10

   (b) Yes. Category (B) conditions exist. 25

   (c) Yes. Category (C) conditions exist. 50

   (d) Yes. Category (D) conditions exist. 65

Score (as
negative values)

(e) Data do not fit any of the above
      categories. Define details below. **30**

(f) Data inadequate to form a
      judgement. Define details below. **50**

Q5. Reply (e)—Atypical nature of gassing data.

....
....
....

Advise Environment Division Director of atypical conditions.

Q5. Reply (f)—Data inadequacies/extra information required.

....
....
....

Advise Environment Division Director that risk evaluation has been biased
because of data inadequacies.

Overall Score   % _____

[C]   Risk reduction achieved by site remediation

Score (as
positive values)

1. Have the remediation works   (a) Yes.   10
   properly focused on preventing/
   reducing on-site gas and/or vapour (b) Not entirely. Specify data below.   5
   risks? (this assumes that the actual (c) No. Specify data below.   0
   gas/vapour source has been
   properly identified).

Note: If the probable source(s) of any gas/vapour hazard has not been properly
         identified, then remediation benefits are unlikely to be adequate.

Q1. Remediation focus on on-site gas/vapour risks.

....
....
....

Score only *one* of the remediation solutions—Q2 to Q5.

*Score (as
positive values)*

2(a). Have the materials, which give rise to the gas/vapour hazards, been entirely removed from the site?

   (a) *Yes.*   70

   (b) *Not entirely. Specify data below.* 35

   (c) *No. Specify data below.* 0

Q2(a).

....

....

....

(b) If such removal has taken place, does quality control or post-remediation monitoring data exist to prove this?

   (a) *Yes. Specify evidence below.* 20

   (b) *Not conclusively. Specify evidence below.* 10

   (c) *No. Specify deficiencies below.* 0

Q2(b). Quality control/monitoring proof.

....

....

....

3.(a) If the materials which give rise to gas vapour hazards have been left on the site, have effective dissipation (venting measures) been included?

   (a) *Yes. Specify details below.* 50

   (b) *Not entirely. Specify details below.* 25

   (c) *No. Dissipation measures seem inadequate. Specify deficiencies below* 0

Q3(a). Adequacy of dissipation methods.

....

....

....

(b) If dissipation measures have been included, is there proof of their effectiveness?

   (a) *Yes. Specify details below.* 20

   (b) *Not conclusive. Specify details below.* 10

   (c) *No. Specify details below.* 0

Q3(b). Quality control/monitoring proof.

....

....

....

4.(a) Have "passive" protection   (a) *Yes. Specify details below.*   40
measures been adopted as the   (b) *To an incomplete extent. Specify*
primary protection method? (e.g.   *details below.*   20
gas-proof membranes, avoidance
of gas traps, installation of   (c) *No. Specify details below.*   0
automatic gas meters, etc.)

Q4.(a) "Passive" protection measures.

....

....

....

(b) Where "passive" measures have   (a) *Yes. Specify details below.*   20
been installed as the primary   (b) *Only to an extent. Specify below.*   10
safeguard, does evidence exist to
prove their effectiveness?   (c) *No. Detail data deficiencies below.*   0

Q4.(b) Effectiveness of "passive" measures.

....

....

....

5.(a) If remediation consisted of a   (a) *Yes. Details and reasons below.*   50
particular mix of the above   (b) *Not entirely. Details and reasons*
techniques, does this appear   *below.*   25
likely to be effective in the long
term?   (c) *No. Details and reasons below.*   0

Q5(a). Mix of remediation and likely effectiveness.

. . . .
. . . .
. . . .

(b) In any such case does quality control and/or post-remediation monitoring proof of effectiveness exist?

(a) Yes. Specify details below.   20

(b) Evidence not conclusive. Specify details below.   10

(c) No. Specify details below.   0

Q5(b). Quality control/monitoring proof.

. . . .
. . . .
. . . .

Overall Score   %   _____

Remnant Risk   %   _____

(Section [B] score – Section [C] score)

9.3   GAS AND/OR VAPOUR RISK EVALUATIONS

The three-part evaluation process described in Section 9.2 differs from other evaluations in that a clear distinction between an initial desk study (part [A]) and evaluation of site investigation data (part [B]) is not as absolute. The mobility of gases and vapours is such that their presence need not be obvious at site surface, if a deeper migration route exists to permit lateral gas/vapour movements. Thus an initial evaluation of site investigation information is required to establish whether the potential for a future gaseous liability exists.

9.3.1   Site examples

Site No. 1

Example Site No. 1 (Appendix I) obviously had had the potential for future landfill gas liabilities, since domestic refuse and ashes were visible and surfaced the land.

The developer's decision to entirely remove these materials, however, dispelled the most apparent landfill gas concern and appeared to make unnecessary any gas measurements or further gassing assessment. However, it is apparent, from List 9.1, that heavily "oil"-soaked ground often gives rise to gaseous hazards, and it is a matter of commonplace observation that car breaking yards usually produce concentrated oil spillages.

Thus it could have been anticipated that difficulties would exist, and that gas surveys should have been undertaken, once the degradable fills had been removed.

Fortunately the opportunity to rectify this oversight and undertake gas surveys was presented, when re-investigation was compelled because of plant die-back (Appendix I). This revealed that a mixture of fuel and lubricating oils and hydraulic fluids had indeed collected in the base of the landfilled gravel pit. Oily fluids existed over a 400 m² plan area and as a floating layer (some 30 mm thick) lying on the groundwater. The water itself contained oily emulsions, and was held up by the underlying glacial clays. Volatile organic vapours (Table 9.4) were measured in boreholes drilled into the old quarry.

Since already occupied homes had been built over the old gravel pit, the option of removing the oily rubble and entirely draining the quarry was not feasible. Residents also rejected any suggestion of retrofitting gas-proofing measures to their homes. Thus the decision was to de-water the area, remove a total of 610 m³ of oils and water which were taken off site, and vent the area of the old pit by installing five stone-filled vent columns at 12 m centres. Post-remediation vapour concentrations are listed in Table 9.4. This remediation, however, did not remove oils held on and in the rubble fills, though these do not constitute other than very small volumes.

Although no UK guidelines exist for organic vapours, concerns are voiced in the USA and Germany that these substances may pose fire and carcinogenic

Table 9.4. Total organic vapour concentrations (mg/litre of air) in infilled gravel pit

Observation borehole	Vapour concentrations on:		
	20 Dec 1991 (pre-remediation)	10 Jan 1992 (pre-remediation)	16 March 1992 (post-remediation)
1	84.0	99.0	12.0
2	115.0	168.0	8.5
3	4.3	<0.1	<0.1
4	19.0	79.0	<0.1

risks. Thus the permissible exposure to coal tar naphtha (a similar solvent mixture) is taken as 100 mg/litre of air, and this value was accepted as applicable.

On the basis of the additional remediation work, the vapour risks which might affect one small group of four homes can be quantified. The probable magnitude of a future liability appeared reasonably high (– 35), since the vapour surveys were not carried out at the time of year (summer) most likely to reveal worst conditions and the surveys themselves were limited in scope, and certainly could not fully predict which gassing conditions might occur in a hot summer (though it is likely that only Category A gassing conditions could occur). The sole positive aspect in this part [B] evaluation is that the actual source of any vapour hazard is certainly known.

Risk reductions by remediation, however, proved to be reasonably adequate.

The remediation work had not been entirely focused on removing all volatile organic hazards since oil soaking of the rubble fill remained, though the proven removal of the floating oils and water mixture did eradicate the major concern. The lack of any longer-term proof, that the venting works were effective, is also an obvious and unfortunate deficiency. However, a remediation benefit score (of + 50 to + 60) would appear to dispel future concerns. In such a case, it would seem prudent, however, to insist on a further organic vapour survey in summer conditions.

Site No. 2

Example Site No. 2 (Appendix I) had a fill capping in which wood and fabric scraps occurred and this gave rise to relatively low levels of landfill gas. These particular fills seemed to extend only to depths of 1.5 m. Lower fills were recorded as containing little or no degradable matter, though it is known that casting sands (because of the presence of phenolic binding agents), which do occur, can generate small volumes of landfill gases.

Gas surveys were included only in the initial site investigation and were omitted from other phases of the site reclamation.

The later re-exploration of the site (compelled by a potential purchaser's concerns that buried contaminated materials could affect the site's clean cover) involved a limited gas survey (Table 9.5), which was expanded in view of the weekly results obtained.

The initial gas concentration surveys (boreholes 1 to 4) had revealed enhanced carbon dioxide concentrations and depleted oxygen values (down to 10% by volume). Since the carbon dioxide levels were in excess of those specified in the Building Regulations (Department of the Environment, 1985), concern was voiced, though methane concentrations invariably were trivial, despite measurements taking place in a warm summer.

This concern led to a series of deeper boreholes (Nos 5 to 10) being constructed to check whether or not the gassing conditions were localised only to the 2.0 m thicknesses of foundry wastes which had been left on site. These holes, however, showed extremely depleted oxygen levels (down to less than 1% by volume) and carbon dioxide concentrations which locally exceeded levels of 20%. The methane concentrations were, however, trivial.

A final deeper set of boreholes (Nos 11 to 15) confirmed the trend. Oxygen concentrations were at the lower limit of detection, whilst carbon dioxide values reached almost 30% in some observation boreholes.

These concentration measurements were continued for several months and revealed that:

● Methane traces in the four shallowest boreholes only occurred in summer conditions when soil temperatures had risen. No explosive levels ever developed, and no positive gas pressures or measurable outflow rates ever were recorded.

Table 9.5. Abstract of gas surveys—Site No. 2

				Weeks					
	1			4			8		
Borehole	O_2	CO_2	CH_4	O_2	CO_2	CH_4	O_2	CO_2	CH_4
1	18	2	0.1	16	3	0.1	17	2.5	0.12
2	19	1.5	—	18	1.5	0.08	16	3.0	0.16
3	14	3.5	0.1	12	4.2	0.18	10	2.4	0.2
4	16	2.5	0.1	13	3.5	0.2	12	4.0	0.2
5		—		3.5	10	—	2.1	12	—
6		—		2.4	12	—	1.4	15	—
7		—		2.8	13	—	1.0	18	0.005
8		—		1.2	16	0.01	0.6	20	—
9		—		1.1	17	—	0.6	19	0.005
10		—		0.8	19	—	0.2	23	—
11		—			—		<0.1	27	—
12		—			—		<0.1	25	—
13		—			—		0.15	23	—
14		—			—		<0.1	23	—
15		—			—		0.12	26	—

Boreholes 1 to 4—4 m deep, screens over lower 2 m.
Boreholes 5 to 10—8 m deep, screens over lower 6 m.
Boreholes 11 to 15—25 m deep, screens over lower 10 m.

● The problem of enhanced carbon dioxide and very reduced oxygen concentrations was controlled by atmospheric conditions. High barometric pressures removed the higher carbon dioxide levels and restored soil oxygen values to levels in excess of 14% by volume.

Further direct measurement of the migration of gases into the enclosed and known volumes of flux boxes (Figure 9.1), buried at depths well below those of any house foundations, revealed that no upward transfer of heavier carbon dioxide took place. Thus it was concluded that the carbon dioxide problem originated in deeper and still-open limestone mine-workings, and that it did not pose a risk to developments on land surface.

A point of interest is that monitoring boreholes invariably are stress relief zones, into which any gases in the surrounding rocks or soils will preferentially migrate. Thus a monitoring borehole will give exaggerated gas concentrations, compared to those in undisturbed ground. Whilst this can be seen as acceptable when lighter gases are concerned, with heavy carbon dioxide (density 1.5 times that of air) monitoring boreholes are likely to act as collecting sumps and give especially unrepresentative results.

In view of the extent of the gas investigation and the agreed and different sources of the methane and the carbon dioxide gases, further remediation was limited to the inclusion of passive venting works (outside of homes) to counteract any feasible risks from the unlikely increase in methane concentrations and flow rates, and ventilation of house foundations to allow for any unlikely rise of carbon dioxide up the site's profile. Both sets of remediation measures were seen as prudent overprovisions, in excess of any compelled by the gas monitoring information.

The future magnitude of any gas liability was separately estimated for the near-surface methane and the deeper carbon dioxide from the old limestone mines. This (Section 9.2, part [B]) revealed low liability scores (-10 for methane risks and -25 for carbon dioxide concerns), since gas surveys had been particularly adequate in extent, were carried out throughout a year and so encompassed any conditions more likely to give rise to worst gassing conditions. Surveys had been comprehensive enough to prove the two separate sources of the two gases and define the gassing categories (i.e. (B) for carbon dioxide and (A) for methane—List 9.4).

Remediation had, however, properly been focused on the gas risk and the mix of additional remediation measures had been more than enough (at a positive total score of $+70$) to remove any future concern over gas risks.

Site No. 3

The third example site (Appendix I) had been expected to display serious oil contamination and the presence of organic vapours had also been anticipated. This was confirmed by the site investigation, which also revealed that benzene

136

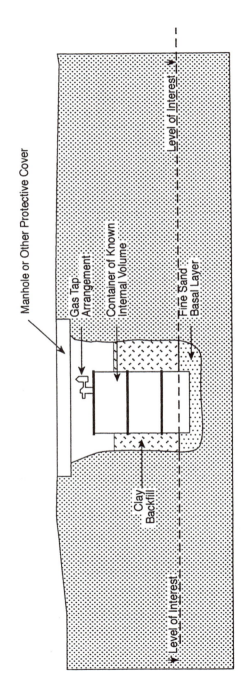

Figure 9.1. Flux box general arrangement

concentrations in the soils were well in excess of allowable exposure levels. Further investigation was undertaken to establish that:

- high carbon dioxide and very reduced oxygen concentrations were also present above the sandstone rock, but
- no abnormal soil atmospheres existed below the oil-contaminated sand layers, which cap the site, and
- no abnormal gases were found to occur in the area of the former gas works.

With the location of gas risks known to be in the upper sand layers and since the health hazards from benzene vapours were properly identified as the greatest gaseous risk, it was reasonably concluded that additional gas surveys would be unnecessary. The remediation sequence described in Appendix I was thus initiated and post-remediation vapour surveys revealed that no gaseous nuisances had persisted and that normal soil atmospheres existed over the reclaimed site. Thus remediation proved to be effective.

This particular site obviously had a high potential for future liability (Section 9.2 [A]) in view of the degree of oil spillages. The full evaluation of the probable magnitude of this future liability (Section 9.2 [B]), however, proved unnecessary, since complete evidence on the sources of the gases of concern was obtained.

9.4 Conclusions

Gas and vapour risks are difficult to interpret, since the measured information can be so very easily influenced by a range of confusing effects. However, the prime aim is always to define the sources of these gaseous hazards so that a provably effective remediation can take place *and* be confirmed. Far too many cases are known where this level of emphasis has not been observed, and too many people have found themselves trapped in homes which now are unsaleable because gas and/or vapour entry has taken place.

10

Risk of Attack on Construction Materials

10.1 MATERIAL DEGRADATION AND FAILURE

10.1.1 Introduction

Concerns over the durability of materials placed in contact with contaminated soil and water have been voiced since the first recognition of the problem of contaminated land. This is understandable. Materials *do* fail in some aggressive conditions, and the consequences can be fatal if load-bearing structural elements collapse.

Material degradation in manufacturing processes is well documented. This has allowed very long lists of substances which can attack materials, in at least some situations, to be drawn up (Barry, 1983; summarised in Table 10.1). Lactic acids (from spilled milk) have slowly dissolved concrete floors in cheese factories, hot sugar solutions in sweet manufacturer's premises have destroyed both floors and vats, and organic acids (such as apple juice) have proved troublesome in some situations.

Lists such as Table 10.1 could suggest that material degradation is near inevitable if the commoner construction materials are exposed to chemical contacts. But are such industrially derived examples really indicative of what can be expected when the exposure is to contaminated land? The answer almost certainly is "No". Chemical concentrations and environmental conditions in manufacturing situations differ far too much from those in contaminated soils for practically useful guidance.

Cases where material attacks have occurred in more natural conditions are, however, more revealing, and indicate the interactions of effects which—together—might lead to the greatest risks of material failure.

A recent interesting study (Henriksen and Stoltzner, 1993) focused on the relatively simple case of salt damage to concrete road bridge columns. The salt, applied to prevent ice formation on road surfaces, forms chloride-rich waters, which then are splashed on to bridge columns at or near the road surface. As

Table 10.1. Compounds which have proved aggressive to common construction materials in manufacturing situations (summarised from Barry, 1983)

	Material	Potentially aggressive compounds
1.	Concrete	56 separate compounds including: fruit products oils sugars tallow various acids various chlorides and chlorine various sulphates vinegar
2.	Steel	63 separate compounds including: beer brines copper and nickel salts milk various acids various chlorides and chlorine various sulphates
3.	Butyl rubbers	18 separate compounds including: aromatic solvents ether mineral oils various acids various nitrates
4.	PVC products	25 separate compounds including: alcohol chlorides and wet chlorine ether and aromatic solvents naphthalene

chloride ions enter the alkaline concrete, they reduce its internal alkalinity and then begin to corrode the steel reinforcement. This attack will usually be apparent on column external surfaces, as the concrete cover above the corroding steel is expanded and cracked. All authoritative guidance (e.g. American Concrete Institute, 1991) warns of such problems and indicates critical chloride ion thresholds (usually at levels of 0.2% chloride ions per weight of cement in the concretes) above which attack is likely. Thus it would be expected that all the bridge columns (in the Henriksen and Stoltzner study) would have shown corrosive effects if their chloride concentrations exceeded such a critical level.

In fact (Table 10.2), many of the examined columns fitted this hypothesis, though the critical chloride ion concentration was rather higher (at 0.30%) than that quoted elsewhere. However, some columns, which were found to hold twice

Table 10.2. Incidence of visible corrosion damage to bridge column reinforcements (after Henriksen and Stoltzner, 1993)

	Chloride concentrations* at ground level				
	0.24	0.3	0.36	0.42	0.48
Percentage of cases where corrosion was visible (Group I structures)	nil	40	100		
Percentage of cases where corrosion was visible (Group II structures)	N/A	N/A	N/A	33	67

*Chloride concentration in terms of percentages per weight of cement in the concretes.
N/A = none apparent.

this critical chloride content, showed *no sign at all* of internal steel corrosion. The investigators found that this was due to:

● better construction workmanship in the unaffected columns,
● thicker and less moisture permeable concrete covers above the steel reinforcement in these cases, and
● better location and drainage works, which reduced the persistence of chloride-rich water contact with the more durable columns.

This study makes a most valid point. No simple material degradation/exposure to chemical compound relationship can be claimed, and other factors have to exist before the potential for material attack becomes a reality. The investigators also made the revealing comment that degradation of materials is both well and poorly understood. Knowledge of the effects caused individually by particular parameters (e.g. the chloride concentrations needed to initiate corrosion) is remarkably detailed and comprehensive, but predicting the likely outcome when several causative factors inter-react is still extremely imprecise.

In some ways this resembles the information situation that is available on gas and vapour emissions (Chapter 9), and gives rise to similar interpretation difficulties.

Other concrete failure cases bring out more clearly the interactions of attack mechanisms which have to occur (Haynes and Zubiate, 1973; Murdock and Brook, 1979). The first of these is the classical demonstration that chloride content (even when that of warm sea water) by itself will have little or no effect on concretes. Sixty-seven-year-old test blocks, which had been submerged off the Los Angeles breakwater, were found when tested to have remained essentially unaffected. In the second case, however, concrete blocks facing the Le Havre sea wall expanded, cracked and collapsed into rubble after only four years of exposure. The differences between these two examples emphasise the salient issues:

● for corrosion of metals to occur both air and water are normally essential, and
● corrosion calls for continual replacement of the aggressive ions, and so alternate wetting and drying is required. If the concretes had been totally

water saturated this could have satisfied the need for chloride replenishment, but at the cost of precluding the essential entry of air.

Exposure to the main aggressive chemical by itself is not enough to cause material attack. Other supportive factors have to occur.

Other cases make the same point. Particularly interesting (Tretheway and Chambers, 1988) was when a stainless steel pipe to carry steam in an industrial plant had to be laid below a road bed. Stainless steel had been chosen as a material unlikely to be attacked by aggressive ground. Despite this, the pipe cracked and failed in only two years. The cause was found to be salts, spread on the road each winter, which ran off the tarmac, percolated into the loose sandy soil on the road verge, and then moved down to collect in the pipe trench. This had had no drainage outflow installed, and so the pipe was exposed to chloride ions at very high concentrations. These were replenished periodically in both winters, but air entry could also occur between wetter periods. Failure in fact resulted from corrosion-induced stress cracking. The salient issues in this case were:

- high chloride concentrations were available for attack,
- regular replenishment of the aggressive chemical occurred,
- easy air entry to support corrosion was possible, and
- poor workmanship was important. Had the trench been properly drained to prevent pools of chloride-rich water collecting, it is doubtful if pipe failure would have taken place.

To further resolve the complexity of the issues involved, it is necessary to consider briefly each of the commoner construction materials in turn.

10.1.2 Concrete

Concrete is by far the most important construction material, and thus has been extensively studied in great detail.

Attacks by sulphates are accepted as the greatest risk to concretes and conditions which make this more likely are known to be as follows (Orchard, 1979):

- When the sulphates present are soluble and reactive. A wide range of different solubilities typifies the commoner sulphates (Table 10.3).
- If concentrations of more soluble sulphates are high. A linear relationship exists and material attack becomes greater as soluble sulphate concentrations increase. However, for continued attack, sulphate ions have to be replenished. This obviously indicates that, if a mobile (moving) groundwater containing sulphates exists, then degradation risks will be greater. Conversely, laying a concrete slab on dry ground (even where this has a high sulphate content) should *not* create problems.

Table 10.3. Some solubilities of
commoner sulphates

Sulphate	Solubility (g/litre)
Calcium	1.44
Magnesium	150
Sodium	200

- If conditions are more acidic and soil or water pHs reach levels of 4.5, or less, degradation will be most apparent. This is because the highly alkaline environment of concretes give stability and protection, which progressively is lost by acid attack.
- If temperatures are higher. As temperatures rise, material attack increases in rate. Whilst this will be unusual in most real-life situations (though it was an obvious factor in the considerable number of industrial failures of concrete floors and vats—Section 10.1.1), it could be relevant if a site is subjected to subterranean smouldering and also to high sulphate contamination. This is not improbable on some old colliery or coke works sites.
- If concretes are more permeable and able to permit easier entry of moisture and the aggressive chemicals. Higher permeabilities arise when more water is added to cement and aggregate mixes than is actually needed for the concrete reactions. All modern codes of practice and design guidances (e.g. American Concrete Institute, 1991; Building Research Establishment (BRE), 1991) thus specify maximum water to cement ratios for different chemical and acidic conditions in soils.
- If the concrete is placed in situations more likely to cause material attack. If all the sides of a concrete block are exposed to water-borne contaminant, attack risks will be less than if water containing high sulphate concentrations exists under a high pressure on one side of a concrete wall or pipe. There the potential for inflow into the concrete will be high and material attack more likely. However, laying a concrete slab so that only the bottom face is in contact with moist sulphate-rich soil gives rise to some risks. A very slow inflow of external moisture to balance evaporation from the upper slab face is likely to occur.

Resistances of various types of concrete to the usual range and concentrations of contaminants which can exist on contaminated sites, is reasonably well known (Table 10.4). This suggests that few contaminated land conditions are likely to be a cause for concern, especially if concretes are designed to modern standards (BRE, 1991) and placed with good workmanship.

Current guidance (BRE, 1991) takes note of the consequences of soluble sulphates being present in various concentrations, whether soil conditions are more or less acidic, and if flowing groundwater exists. The five different classes of concrete which are today specified (BRE, 1991) allow for the feasible range of

Table 10.4. Resistances of concretes to degradation by various substances (after British Standards Institution, 1971; Orchard, 1979)

Little or no degradation	Mild degradation	Heavy degradation
(a) *Portland cement concretes*		
Alkalis (<10% solutions)	Acidic waters (natural)	Strong acids
Brines	Chlorides (if air present)	Strong hydroxides
Coal tar constituents	Cinders and ashes (if wet)	Sulphates
Oils (mineral and petroleum)	Creosotes	Vegetable oils
Organic solvents		
(b) *High alumina cement concretes*		
Carbonic acid	Acids if pH >4	Strong acids
Creosotes	Alkali hydroxides	
Phenols	Sulphur dioxide gas (if moist)	
Sulphates		Strong alkalis

conditions which can be encountered. If such guidance is properly followed, it is very difficult to believe that concrete degradation should occur.

A cautionary note is, however, essential. Some very usual contaminated land situations, such as the ash surfaced sites which are so commonplace, pose few difficulties when the ground is dry. Soluble sulphate concentrations can be high (Table 10.5), but in the lack of a water transport mechanism this is of little consequence. However, the same ashes and cinders if water saturated (by, for example, a particularly high annual rise in groundwater level) can become distinctly more acidic and also are more able to penetrate concretes. Thus it will always be necessary to predict which worst-case conditions can arise, and not merely apply the available guidance in a mechanistic manner.

10.1.3 Steel

Steel is also a commonly used construction material, employed mainly as reinforcement to improve concrete resistance to tension cracking.

Despite examples such as the Le Havre breakwater and the bridge column studies mentioned earlier, failures of reinforced foundation slabs or piles are rare (BRE, 1991). This is because of the different exposure circumstances. An examination of the processes involved in the corrosion of steel makes the point clear:

- Concrete is very alkaline (pHs of about 12.5).
- In such conditions, steels become coated with iron oxides. These not only prevent any further rusting, but remain self-healing as long as the alkalinity is high.
- If acidic salts or solutions can enter the solid concrete, these cause the alkalinity to fall.

Table 10.5. Aggressive content of a typical ash-covered site (West Midlands)

Trial pit	Total sulphate content, mg/kg (acid extract)	Soluble sulphate content, mg/kg (2:1, water:soil extract)
1	3 300	820
3	472	105
7	13 000	2 161
9	29 000	14 400*
12	11 000	3 600
19	200	18

*Abnormally high content of sodium sulphates proved to be present.
Conditions of site: Ash layer 2.5 m thick; foundations in ash layer. Groundwater level (August 1992) at 4.3 m; groundwater level (March 1993) at 1.5 m. Sulphate concentrations in 1993 groundwaters averaged 861 mg/litre and attained a known peak value of 1892 mg/litre. pH of March 1993 groundwaters 5.1.

- At critical lower pHs (of about 10) the protective iron oxide coatings fail, and at this stage it first becomes possible for corrosion of steel to commence.
- For this potential to become a reality it is necessary that *both* air and water are able to enter the solid concrete.

Thus, three essential conditions have to exist:

- aggressive and acidic compounds have to occur at high enough concentrations in the soils, or groundwaters, in contact with the concrete,
- the concrete has to be permeable enough to allow the entry of solutions of these compounds, and
- water and air also have to be able to penetrate into the solid concrete.

In most cases typical of contaminated land conditions it will be improbable that these conditions will be fulfilled.

(a) Chlorides, sulphates and carbon dioxide (from the atmosphere) are all able to reduce concrete alkalinity and initiate corrosion processes. Of these, chloride ions are accepted as the most hazardous. Guidance codes (e.g. American Concrete Institute, 1991) thus specify maximum allowable chloride ion concentrations (usually at levels of about 0.2% chloride ions per mass of cement in reinforced concrete and 0.08% chloride ions per mass of cement in pre-stressed concrete). This type of guidance refers to the *total* chloride content in a concrete and not to any safe level in the environment around the solid concrete, as chlorides can occur in the sand, gravel, water and various accelerators (to speed up concrete setting times) which formed the original mixes. High chloride contents in the materials used to manufacture concrete are especially likely to occur in the saline conditions of the Middle East, and studies (Matta, 1992) there have indicated that the safe external chloride concentrations will be (for reinforced concrete) about 0.1% chloride ions per mass of cement used. This information offers a choice of solutions to achieve concretes which should not contain overly high aggressive compounds. Either:

- 0.1% chloride ion content can be taken as the maximum allowable levels in soils or groundwaters which surround the concretes, or
- 0.2% chloride ion content can be substituted, but only if it is known that saline sands, gravel, waters and setting agents will not be used in concrete manufacture.

(b) For any aggressive compounds to penetrate hardened concrete to attack steel reinforcement, the concrete has to be more permeable, or the concrete cover laid over the steel has to be too thin. Neither condition should arise if proper note is taken of concrete guidance codes (e.g. American Concrete Institute, 1991; BRE, 1991). Permeability in concretes is largely a function of the water content used to slake the cement in the original concrete mix. Too high a water : cement ratio will give more permeable concrete. However, codes of guidance do specify maximum allowable water : cement ratios. Likewise concrete cover thicknesses, over steels, in the range of 38 to 50 mm should be adequate to prevent meaningful inflow of aggressive solutions, and problems tend to occur only when thinner covers (about 12 mm or so) are employed. A final factor which would encourage the entry of aggressive solutions would be if the hardened concretes had been poorly constructed and had cracked and split. This, however, should not be a point to concern environmental risk assessors. A range of other professionals—e.g. architects, resident site engineers, and building control inspectors acting for local authorities—will have already given concrete construction workmanship and concrete durability a very high priority, and will have checked the quality of concrete members, structures and foundations at regular intervals during constructions.

(c) The final condition—that water and air both have to penetrate solid concrete to accelerate corrosion—is difficult to accept in most sub-surface conditions. If the ground is water saturated, air is very unlikely to be available (not only do oxygen contents decline rapidly with depth in a soil profile but air entry through water-filled pores will—at best—be minuscule).

(d) Reinforced concretes are generally employed on reclaimed and previously contaminated sites as foundations, sub-surface service pipes and drains. In these situations, exposure to corrosion effects is markedly reduced. If—as should be the case—workmanship in constructing and placing the concretes has been adequate (a point regularly checked by other professionals), failures of reinforced concretes should be improbable. Enough information exists to establish the levels of external chloride contents likely to be of concern and these (at 0.1%) are far above the conditions known to occur on most contaminated sites. Even should such unlikely high chloride concentrations exist, the usual lack of oxygen at a few metres depth (about 2.0 m in clay soils) in sub-surface conditions removes an essential element for corrosion to be significant.

Where steels or cast irons are used without protective concrete coverings (usually as pipes), corrosion is more likely. This will be particularly the case where soils are acidic, and aggressive chemicals (sulphates and chlorides) co-exist with available water and air.

Despite this enhanced susceptibility, good workmanship (e.g. including adequate drainage measures, back-filling with alkaline limestone gravel, etc.) can markedly reduce corrosion potential. Such measures will usually be demanded by architects and building control inspectors.

One unusual condition has also to be noted. Sulphate-reducing bacteria can (in the absence of soil oxygen) convert sulphates to sulphides in very acidic ground (pH < 4). These sulphides can then corrode steels and cast irons. This situation can occur in old gas and coke works and in areas where sulphate-rich wastes (e.g. plaster wastes) have been tipped.

10.1.4 Plastics and rubbers

Whilst a large body of information exists on the conditions likely to degrade traditional construction materials, and codes of practice allow necessary levels of good workmanship to be specified, the same cannot be said about a wide range of newer synthetic products. These include the various plastics and rubbers.

None of these has been in use long enough for durabilities to be known with confidence, and manufacturing variabilities are such that marked susceptibility differences are likely to exist.

Thus, it will be prudent to check with manufacturer test data before synthetic sheeting, reinforcement, sealants and pipes are chosen for use in contaminated soils or groundwaters.

Many of these newer products are prone to attack from mobile organic compounds (e.g. phenols in particular), strong acids or alkalis, higher soil temperatures and some microbial conditions. Others (such as the synthetic rubbers) may be susceptible if exposed to catalytic oxidation (caused by—for example—high copper concentrations) or to swelling and failure (if soaked by oils and organic solvents).

10.1.5 Good workmanship

The European-wide economic recession of the early 1990s led to the closure and demolition of many manufacturing and business premises. Where these had been constructed in the 1960s and 1970s, with modern construction materials, but also in contaminated soils, it is revealing that the demolished materials still remain in sound condition. Material attack usually can only be seen where poor quality workmanship had taken place.

This practical fact (which is confirmed by demolition contractors particularly active in this type of work) demonstrates that, where materials are selected in line with guidance codes and where workmanship is of a normal standard, material durability is adequate to resist most contaminated land conditions. Thus undue concerns of risks to construction materials seem not to be well based.

10.2 MATERIALS ATTACK RISK ASSESSMENT

Main Concerns

That concrete, steel, iron, plastic or rubber components could be degraded by contamination contact, and create damage to, and ultimate failure of, structures. Physical failures (e.g. subsidence) would also be a concern.

These concerns will be of greatest importance to individual house owners, and will also be of interest to local authority building control officers.

The presence of a limited range of contaminants (List 10.1) in mobile/soluble forms will make the concerns greater.

For these concerns the worst-case assumption that sensitive targets exist has to be made.

[A] Potential for future liability

Tick appropriate box

1. Does a fill layer mantle the site, and is it likely to extend to the depths of normal foundations?

    (a) *Yes. Fill covers > 50% of site area.* ☐

    (b) *Yes. Fill extent < 50% of site area.* ☐

    (c) *No. Site appears to lack surface fills.* ☐

2. If a fill exists, is it likely to contain List 10.1 contaminants in high concentrations?

    (a) *Yes. Past site use would be expected to generate these materials at high concentrations.* ☐

    (b) *No. List 10.1 materials not likely to be present at high concentrations.* ☐

3. If List 10.1 contaminants are likely to occur, would these be expected to be in mobile/soluble states? (note likely origins, local ground-water quality, etc.)

    (a) *Yes. Specify evidence below.* ☐

    (b) *No. Specify evidence below.* ☐

Q3. Evidence for contaminant mobility/solubility.

. . . .

. . . .

. . . .

4. If List 10.1 contaminants are likely to occur, is it probable that a near-surface (<3 m) groundwater could be a transport mechanism?

   (a) *Yes. Groundwater likely to be near surface.* ☐

   (b) *No. Groundwater not likely to be near surface.* ☐

5. Are soil or near-surface ground-waters likely to prove acidic?

   (a) *Yes, acidic soils or waters likely to have resulted from site's past uses.* ☐

   (b) *No. Past site uses unlikely to have given rise to acidic conditions.* ☐

6. If no fill layer exists, is it likely that spillages of List 10.1 substances could have occurred on the site to a significant extent?

   (a) *Yes past land uses indicate tanks of such substances existed.* ☐

   (b) *No. No such evidence exists.* ☐

7. Is there evidence that the site has already degraded construction materials? (examine old foundations, etc.)

   (a) *Yes. Old concrete/iron/steel heavily attacked. Specify evidence below.* ☐

   (b) *No. Evidence for such attack not visible.* ☐

Q7. Evidence for existing attack on construction materials.

....

....

....

---

8. Is there any likelihood of unstable ground conditions because of subterranean smouldering or very poor compaction of fill layers?

   (a) *Yes.* ☐

   (b) *No.* ☐

If answers to all the above questions are "No", and if an initial review of available Site Investigations data confirms this, terminate construction material attack liability assessment at this point.

    For other results, summarise the overall evaluation below. Justify clearly any decision to terminate construction material attack liability assessment.

....

....

....

....

....

LIST 10.1

Contaminant conditions which may pose risks to the commoner construction materials

	Material	Contaminant conditions of concern
1.	Concrete	If pH low (<4.5) If mobile groundwater exists If soluble sulphates >1.2 g/litre If chloride and nitrate contents also high If inadequate note has been taken of Building Research Establishment (BRE) (1991) guidance
2.	Reinforced concrete	If pH is low (<4.5) If air and water occur If chloride content (external) is >0.1% per mass of cement If inadequate note has been taken of guidance (American Concrete Institute, 1991; BRE, 1991) on cover thickness, water:cement ratios, and workmanship standards
3.	Steels and cast irons	If pH low (<5) If air and water are present If sulphate-rich wastes occur at very low pHs (<4) If sulphides exist (>1000 mg/kg)
4.	Synthetic materials	If mobile organic fluids (particularly phenols) occur (>200 mg/kg) If free oils or solvents exist (seek manufacturer test guidance) If enhanced soil temperatures exist (>20 °C) If organic wastes occur If high copper concentrations occur (seek manufacturer test guidance)

Scoring Options	**Score**
1. Where, for any material*, three or more of the listed concerns exist.	30
2. As above, but only two of the listed concerns exist. One of these should be soil pH.	20
3. As above, but only one of the listed concerns exists.	10
4. As above, but none of the listed concerns exist.	0

(Materials in question will be those used in foundations and buried pipes and services.)

[B] Probable magnitude of a future liability

*Score
(as negative
values)*

1. Do high concentrations of the List 10.1 substances exist on or in the site and at least to normal foundation depths?

    (a) *Yes. Present at high concentrations. Specify details below.*   10

    (b) *To an extent. Specify details below.*   5

    (c) *No. These contaminants absent, or present only at low concentrations. Specify details below.*   0

Q1. Presence of List 10.1 substances.

....

....

....

---

2. If these substances are present, are they in mobile/water soluble forms? (note origins and groundwater evidence, etc., and see end of List 10.1 for scoring options)

    (a) *Yes. Specify evidence below.*   30–20

    (b) *To an extent. Specify evidence below.*   20–10

    (c) *No. Evidence (below) is that substances are immobile/non-soluble/of low solubility.*   10–0

Q2. Mobility/solubility proof.

....

....

....

---

3. Does a local transport mechanism (e.g. a near-surface groundwater) exist to encourage upward movements of List 10.1 substances (loading of mobile substances also of significance) to attack foundations and materials?

    (a) *Yes. Specify details below. (e.g. groundwater < 3 m depths).*   20

    (b) *To an extent. Specify details below. (e.g. groundwater > 5 m depths).*   10

    (c) *No. Foundation depth will not be affected by deeper contaminants. Specify details below.*   0

Q3. Local transport mechanisms.

....

....

....

Score
(as negative
values)

4. Are the planned depths of foundations and buried services known and within contaminated materials?

(a) *Yes. Foundations and services will be within contaminated materials.*   10

(b) *Whilst some are founded in clean soil, others do have to lie in contaminated material.*   5

(c) *Foundations and services locations unknown.*   5

(d) *All foundations and services will be located in clean materials.*   0

5. Is site subsidence a meaningful risk on this site?

(a) *Yes. Materials particularly loose/ill-compacted/liable to settle (e.g. peaty soils).*   10

(b) *To an extent. Compaction state intermediate/some thinner compressible materials do exist. Specify details below.*   5

(c) *No. Materials well compacted/with high bearing capacities/lacking compressible bands.*   0

Q5. Evidence for possible subsidence.

. . . .
. . . .
. . . .

---

6. Could the site be affected by subterranean fire/smouldering? (relate to Chapter 9 evaluation)

(a) *Yes. High content of combustible fill/materials. Specify details below.*   10

(b) *To an extent. Some areas under-lain by potentially combustible fills/materials do exist. Specify details below.*   5

(c) *No. Specify evidence below.*   0

Q6. Evidence for subterranean fire/smouldering potential.

. . . .
. . . .
. . . .

7. Is the Site Investigation adequate   *(a) Yes.*   0
   to permit the above questions to be
   answered with confidence?   *(b) To an extent. Specify deficiencies   5
                                              below.*

                                        (c) No. Specify deficiencies below.   10

Q7. Information deficiencies.

. . . .

. . . .

. . . .

Advise Environment Division Director if risk evaluation has been biased because
of data inadequacies.

<u>Overall Score</u>   %   _____

[C] Risk reduction achieved by site remediation

1. Has the remediation properly   *(a) Yes.*   10
   focused on preventing/reducing
   risks to construction materials and *(b) Not entirely. Specify data below.*   5
   structures?   *(c) No. Specify details below.*   0

Note: If the probable risks to materials and structures have not been properly
identified, then remediation benefits may not be entirely adequate.

Q1. Remediation focus adequacy.

. . . .

. . . .

. . . .

Score only *one* of the remediation solutions—Q2 to Q5.

2.(a) Have the substances which   *(a)*   *Yes. Provide details below.*   **70**
could attack materials *and/or*
support subterranean fires/   *(b)*   *Not entirely. Specify data below.*   **35**
smouldering all been removed   *(c)*   *No. Specify data below.*   **0**
from the site, and has the
required site bearing capacity
then been achieved?

Q2(a).

. . . .

. . . .

. . . .

(b)   If such removal has taken place,   *(a)*   *Yes. Specify evidence below.*   **20**
does quality control and/or post-
remediation monitoring data   *(b)*   *Not conclusively. Specify evidence*
exist to prove this and that   *below.*   **10**
required bearing capacity now   *(c)*   *No. Specify deficiencies below.*   **0**
exists?

Q2(b). Quality control/monitoring proof.

. . . .

. . . .

. . . .

3.(a) If substances which could attack   *(a)*   *Yes. Cover/clean replacement fill/*
materials and/or support sub-   *immobilised layer is a structurally*
terranean fires/smouldering have   *sound platform which is not likely to*
been left on site (but below the   *be recontaminated by rising*
levels of foundations and   *contaminants* **or** *permit air entry to*
services), is the reclamation   *potentially combustible materials.*
likely to prove effective and   *(Have this situation checked by*
permanent?   *clean cover methodology and/or*
        air entry criteria through compacted
        materials.)   **50**

    (b)   *To an extent. Specify details below.*   **25**

    (c)   *No. Capping/clean replacement fill*
        seems inadequate structurally/ as
        a break layer/as an air entry barrier
        (delete as appropriate). Specify
        details below.   **0**

Q3(a). Adequacy of cover/clean replacement fill.

. . . .
. . . .
. . . .

(b) If such a cover/clean replacement fill had been included, is there proof of its effectiveness?

    (a) *Yes. Specify details below (note three distinct functions which may be required).* 20

    (b) *Not conclusive. Specify details below (note three distinct functions required).* 10

    (c) *No. Specify details below (note three distinct functions required).* 0

Q3(b). Quality control/monitoring proof.

. . . .
. . . .
. . . .

4.(a) Have "passive" protection measures been adopted as the primary protection method (e.g. protection of materials, selection of less easily attacked materials, etc)?

    (a) *Yes. Specify details below.* 40

    (b) *To an incomplete extent. Specify details below.* 20

    (c) *No. Specify details below.* 0

Q4(a). "Passive" protection methods.

. . . .
. . . .
. . . .

(b) Where "passive" protection methods have been installed as the primary safeguard does evidence exist to prove their effectiveness?

    (a) *Yes. Specify evidence below.* 20

    (b) *Only to an extent. Specify below.* 10

    (c) *No. Data deficiencies below.* 0

Score
(as positive
values)

Q4(b). Effectiveness of "passive" measures.

. . . .

. . . .

. . . .

5.(a) If remediation consisted of a particular mix of the above techniques, does this appear to be effective in the long term?

   (a) Yes. Details and reasons below.   **50**

   (b) Not entirely. Details and reasons below.   **25**

   (c) No. Details and reasons below.   **0**

Q5(a). Mix of remediation and likely effectiveness.

. . . .

. . . .

. . . .

(b) In any such case does quality control and/or post-remediation proof of effectiveness exist?

   (a) Yes. Specify details below.   **20**

   (b) Evidence not conclusive. Specify details below.   **10**

   (c) No. Specify details below.   **0**

Q5(b). Quality control/monitoring proof.

. . . .

. . . .

. . . .

Overall Score   % _____

Remnant Risk   % _____

(Section [B] score – Section [C] score)

10.3   RISKS OF ATTACK ON CONSTRUCTION MATERIALS

The three example sites (Appendix I) all had some potential (pre-reclamation) to affect commoner construction materials.

Of these, the oil storage depot site (Site No. 3) was obviously worst. Near-surface groundwaters, which could be very acidic and also contain high sulphate concentrations, occurred over the old gasworks area. Elsewhere high organic concentrations were present. Despite these original conditions, site investigation and remediation had been entirely adequate. Thus it can be claimed that no remnant liability risks will remain.

Whilst the two other sites had been less obviously contaminated, in both cases deficiencies in the information available prevents the same confidence being expressed.

On the former iron foundry (Site No. 2), the lack of any information on the near-surface groundwater's chemical quality is a serious information deficiency. This led to a relatively high magnitude for future liability (-40) being awarded. This could not be fully dispelled when remediation benefits were assessed, since the effectiveness of the site's clean cover will depend on the quality of any contaminated groundwater which lies below it (Cairney, 1993b), and so long-term safety had to depend on the selections of construction materials. Obviously this remnant doubt—which probably is *not* well founded, since iron foundry wastes seldom display high enough solubilities to create high levels of groundwater contamination—could easily have been resolved had additional site investigation information been made available.

On the light industrial site (Site No. 1) no problems of materials degradation should have been predictable, since the developer did remove all surface contaminated fills.

However, the lack of information on the chemical quality of the glacial clays (which were the foundations for houses), *and* the knowledge that very acidic conditions exist where timber treatment liquid spillages had occurred, gave rise to difficulties. It thus could *not* be safely concluded that aggressive chemicals had not collected on or in the upper layers of the clays to give rise to long-term material attack.

These example sites make the obvious point that if information of importance is not obtained, then risks cannot be fully assessed. In the context of risks to materials, this is unfortunate, since these risks will normally be of very limited significance if proper attention is paid to available guidance and codes of practice.

11

Risks to Plant Populations

11.1 VARIATIONS IN CONTAMINATION EFFECTS ON PLANTS

Whilst there is no doubt that contaminated land can adversely affect vegetation (e.g. John and Van Learhoven, 1976; Peterson *et al.*, 1981), predicting whether ill-effects will be significant is not at all easy.

This, in part, is because a range of factors (which themselves can inter-react) control the scale of possible consequences—i.e.:

(a) the types of contaminants at plant root zones, *and* if these occur in plant-available forms,
(b) the soil type, and its acidity or alkalinity,
(c) the density of the soil, which controls the growth and extension possible for plant roots,
(d) whether aeration levels at plant root zones are adequate, and
(e) whether other stresses (such as nutrient deficiencies, lack of available moisture, etc.) also occur to enhance the effects of contaminant attack.

An additional complexity is that plants, like all higher life forms, seldom display simple cause and effect responses. Some species prove to be less susceptible to the presence of contaminants, and, within a single species or cultivar group, individual plants can often exhibit very different reactions to those of their neighbours.

Thus the effects of land contamination on plant life are especially complex, and still are very far from being fully understood.

11.2 PLANTS DIRECTLY ESTABLISHED INTO CONTAMINATED SOIL

The most obvious ill-effects will be apparent if plants are rooted directly into soils which contain plant-available contaminants. However, even in such conditions, variability will occur.

Conditions will prove to be more harmful if a range of metallic contaminants (i.e. copper, nickel, zinc and cadmium) are present, *and* if soil conditions are distinctly acidic. This is because metallic contaminants are most plant-available at low pH levels.

The sensitivity imposed by pH changes can be very striking. Trials (Alloway and Morgan, 1986), for example, have shown that 70% of a soil's cadmium will be soluble and plant-available at pHs between 3.0 and 4.0. When, however, soils become more neutral (pHs of 5.0 to 6.0) the amount of available cadmium falls to only 7%.

Such acidity variations can occur quite naturally and as a result of rainfall leaching. Studies on undisturbed copper smelting waste tips, in the Swansea Valley (Bridges *et al.*, 1979), recorded the expected acidic surface soil conditions, caused by the presence of decomposing plant litter in the surface layers. With depth, however, the soils became increasingly alkaline. The consequences were that the copper present was plant-available to 400 mm depths (to where pHs had risen to 4.81 ± 0.15), whilst zinc was plant-available to much greater depths (i.e. to 800 mm where pHs reached 5.48 ± 0.39). This type of study makes it obvious that no unique acidity level in soil will be critical, and that the contaminants most likely to harm plant life remain soluble and available to quite different points.

Soil acidity, however, is not an independent variable, but is affected by other effects. Increasing a soil's organic content, by, for example, adding shredded bark, sewage sludge or compost, will tend to increase acidity. Thus if phytotoxic contaminants exist, these would be expected to become more plant-available.

Whilst this is usually so, there are cases (e.g. Alloway and Morgan, 1986) where the reverse has been shown. The addition of sewage sludge to various soils, in this study, actually reduced the amount of plant-available cadmium, though not the availability of other contaminants (such as nickel). The difficulty is that quite different contaminant species (Chapter 1, Section 1.3) can result as organic complexes develop in soils.

Whilst these types of confusions can always be resolved (if detailed research is directed at a particular site's conditions), in the generality of routine reclamation of contaminated land, estimation of the availability of contaminants likely to harm plant life presents very real difficulties.

The current UK guidance (ICRCL, 1987) takes note of the problem by indicating critical concentrations of harmful contaminants *only* for the likely worst-case conditions (i.e. in thin, sandy and acidic soils). The guidance note specifically advises that conditions harmful to plant life are unlikely, at the stated contaminant concentrations, if soils are other than sandy and acidic. However, no attention is given to the importance of a soil's organic content or organic type, since this extra complexity proved too difficult to include in a simplified guidance note.

Thus specialists involved in contaminated land reclamations have to operate with only very partial guidance and with considerable uncertainties.

Obviously there are situations where harm to plant life can be predicted with confidence. These generally are where other stresses exist—i.e.:

(a) poor water retention in very loose granular soils,
(b) waterlogging in heavy ill-drained soils,
(c) inadequate aeration at plant root zones,
(d) the presence of gases which can drive out soil oxygen or be directly toxic to plants,
(e) too high a soil compaction to permit root growth and extension and so plant development,
(f) extremes of acidity or alkalinity in soils, and
(g) nutrient deficiencies.

These supplement the ill-effects of contaminated soils and create extremely inhospitable and harmful environments for plants. Such adverse conditions commonly cause large-scale plant cover failure on unreclaimed sites.

Around the toes of colliery spoil heaps (where acidic and metal-rich seepages combine with the ill-effects created by nutrient-poor and overly compacted colliery spoil materials), or on areas of former chemical works (where water soluble contaminants at high concentrations can occur in very loose, easily drained and nutrient-poor demolition rubbles and fills), a total absence of plant life is more the norm than the exception. Likewise, where localised escapes of landfill gases occur, large-scale plant die-off is usual, since soil oxygen levels have been reduced below the minimum needed to support vegetative life.

However, these extreme situations should never be typical of reclaimed sites, unless a total disregard for normal professional standards has been the case. More localised areas of plant deaths do occur on a few reclaimed sites (e.g. Site No. 1, Appendix I) and certainly prove that reclamation there has been inadequate. This, however, is not the greatest of concerns, since plants can be replaced (after the reclamation failure has been corrected) at no great cost and without long-term risk to people or other targets.

Far more serious are those situations where sub-phytotoxic effects occur, as plants draw up available contaminants and concentrate these in fruit and leaves.

Plant response to the presence of available contaminants in soils is known to be both species variable and also influenced by the tolerances of individual plants. A large range of the commoner garden crops has been studied, and trials (e.g. Lepp and Harris, 1980) have shown clearly that whilst some individuals die off and others survive only as stunted and deformed plants (which would be very unattractive for eating), some can appear to be entirely normal and healthy. The problem is that these apparently very edible crops can contain high contaminant concentrations in their tissues. Thus they potentially present direct risks to human health, if taken into the food chain.

Because such sub-phytotoxic responses to being grown in contaminated soil occur so often, detailed studies (e.g. Barltrop, 1986; Thornton, 1986) have been directed at establishing the order of health hazards likely to result. These studies, although proving that contaminant uptake and concentration does occur, have failed to demonstrate any actual human health hazard. This could be because the concerns over eating contaminated crops are exaggerated, or it might be due to much of the contamination being concentrated in the portions of plants which are not usually eaten (e.g. potato skins, stalks of cabbages, etc.). From a practical viewpoint it really does not matter which suggestion is correct; plants are known to be able to concentrate harmful contaminants and people can eat such crops. Thus a level of concern has to exist whether or not we actually are able to quantify the resultant health impact. Human health risks could take a good many years to become obvious, and the effects of eating contaminated crops could be obscured by other contaminant exposure events.

When plants are established directly into contaminated soil, it will often be difficult to predict if any contaminant risk will occur. The available information is far from complete, factors which do influence the risks of plant harm (such as soil acidity levels) can alter, or be influenced by other changes to the soils, and current guidelines are extremely tentative. Thus it will usually be more prudent to provide a clean stratum of surface soil if vegetation cover is needed on areas of a reclaimed site.

11.3 PROVIDING CLEAN SOIL LAYERS FOR PLANTS

If a clean layer of soil is necessary because the existing soils contain unacceptably high concentrations of contaminants (List 11.1, Section 11.4), two requirements have to be fulfilled:

(a) the soil thickness and type have to be adequate for the vegetation that is planned, and
(b) no re-contamination of the soil must be possible from soils, fills and groundwater deeper in the site.

Quite obviously, the thicker an imported soil layer has to be, the greater will be the costs and the problems. Clean soil is not always easy to obtain. Developers, quite naturally, will also seek to minimise the costs of importing soil, and will argue for as thin a cover as possible.

Although the thickness and the type of clean soil are interrelated factors, which together determine if vegetation will survive and prosper, these two aspects are discussed separately (below) for clarity.

The available literature contains a good many references to remarkably deep plant root penetrations, though the factors which could be responsible for these are often not described (Department of the Environment, 1986; Parry and Bell, 1987b). It is not uncommon (for example) to read that grass roots might penetrate to 500 mm depths even in a wet summer, and to more than 1200 mm in drought conditions (Parry and Bell, 1987b). But in what types of soil? With what nutrient and water retention characteristics? With what bulk density?

Without this information it is unclear if such root penetration depths are normal or distinctly atypical.

There also is a widespread belief in deep tap roots—which are pictured rather as biological pile drivers, able to smash their way down to great depths—and in an "hour glass" plant model (Figure 11.1a), in which the depth and extent of plant roots mirror the above-ground development of branches and stems. These older views, however, are no more than misrepresentations.

Plant architecture (Dobson and Moffat, 1993) is now known to be more of a "wine glass" model (Figure 11.1b), and plants tend to have relatively shallow root mats influenced more by soil conditions (density and available nutrients and moisture) than by any species categorisation. Indeed the simplicity of the concept of "deep" and "shallow" rooted plants is now known to be misleading.

There are actual constraints on root growth and extension and, of these, by far the most important are the densities of the soils. At certain critical densities plant root growth is progressively reduced, at others it is entirely halted (Table 11.1). That this information has been known for decades, and yet has been entirely ignored in the context of most land reclamation, is somewhat surprising.

The practical consequences of this recently summarised information (Dobson and Moffat, 1993) is that a particular and suitable thickness of clean soil can be provided for the chosen plant populations (Table 11.2) *and* that deeper root penetration can be prevented by suitably compacting lower levels in the ground profile.

Such compaction (to the densities cited for preventing root growth in Table 11.1) is a routine necessity, to provide adequate ground-bearing capacities (for roads, buildings and houses). Thus the previously held views, on the difficulties of providing a thickness of clean soil below which plant roots will not migrate, now can be discounted.

The thickness of the clean soil layer that is provided is, however, merely one facet of the requirements. Equally, if not more important, is to ensure that the provided type of soil satisfies plant requirements. This calls for soils to have a range of characteristics, and experience (Beauchamp, 1993) has shown that these are satisfied by the limitations noted in Table 11.3.

Where available soil, or soil-forming materials, fail to meet some of these requirements, it can be possible to amend the soil properties (e.g. by adding organic material or fertilizers) (Beauchamp, 1993).

What is clear is that the type of soil provided is crucial if deeper plant root migration is to be prevented. Plant roots do not constantly develop, but extend only to locate the moisture and nutrients they need. Thus, if care is taken to provide a suitable soil layer, plant root extension should not, in fact, occur. Simply, the plants have no need to waste energy if they already have access to their needs.

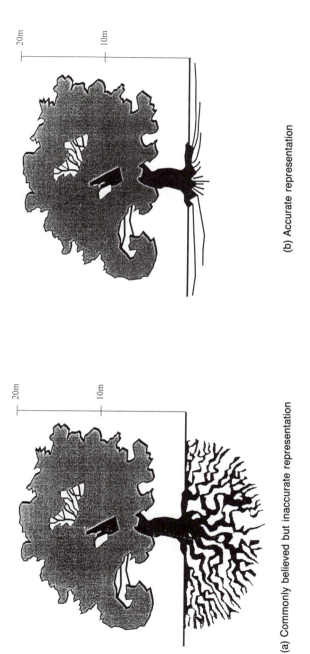

20m

10m

(a) Commonly believed but inaccurate representation

20m

10m

(b) Accurate representation

Figure 11.1. Tree root architecture

Table 11.1. Mechanical resistance to root growth (Heilman, 1981)

Soil density (g/cm³)	Effect on root growth
1.37	Root growth begins to be affected
1.37 to 1.77	Root growth diminishes linearly
1.74 to 1.83	Root growth ceases entirely
1.55 (clay soils)	Root growth severely impeded
1.85 (sandy soils)	Root growth severely impeded

Note: Properly compacted reclaimed site surfaces will exhibit densities above those which will permit plant root intrusion.

A final concern, of clean imported soil becoming contaminated via upward movement from buried contamination, has to be considered. There is, of course, no practical sense in going to the expense and trouble of establishing the right thickness and quality of clean soil just to let this be affected by contaminant uptake.

Upward migration of contaminants usually will be because a near-surface contaminated groundwater exists within 3 m of a site's surface (Cairney and Sharrock, 1993). This effect is most apparent in long hot summers, when the site surface is dried out, and moister conditions still exist at depth in the soil profile. Quantifying the likely magnitude of uptake is not difficult (Cairney and Sharrock, 1993), and—if this seems likely to pose problems—the situation can be rectified. Either the level to which the groundwater can rise can be restrained (by drainage works), or a capillary break layer can be installed to reduce uptake to acceptably low concentrations.

An additional effect, which should not be overlooked, is the possibility that gases could rise up the profile to either expel oxygen (and so asphyxiate plants) or induce toxicity effects. If adequate site remediation has taken place (see Chapter 9), such upward movements of gases and vapours should not usually be a risk.

Providing a layer of clean soil which will adequately provide for plant needs and prevent deeper root migration is not especially difficult. The critical factors are to ensure that lower layers are compacted to provide a mechanical resistance great enough to halt root penetration, and that the rooting soil itself has the

Table 11.2. Minimum soil layer thicknesses for chosen vegetation types (from Heilman, 1981; Beauchamp, 1993)

Vegetation	Soil layer thickness
Grass cover only	150 mm
Garden crops and grass	200 to 300 mm
Shrubs	500 mm
Fruit and specimen trees	1.0 to 1.5 m

Assumptions: 1. That 30% of provided thickness will be of good quality topsoil. 2. That water retention will be adequate.

Table 11.3. Desirable characteristics for clean soil layers (assuming these support normal garden vegetation)

Bulk density	$< 1.5 \, \text{g/cm}^3$
Stoniness	$< 30\%$ by volume. None in excess of 50 mm in size
pH	5 to 8
Water retention	Similar to that of a loam to a clay loam (Beauchamp, 1993)
Organic content	$\sim 10\%$
Nutrient content	Equivalent to good topsoil
Contaminant levels (both metallic and organics)	Not in excess of ICRCL (1987) guidelines for gardens

characteristics required. The sole additional check is always to confirm that the clean soil will not become contaminated from below.

Rooting plants directly into contaminated soil (if contamination levels are not too far in excess of the ICRCL guidelines) could be successful, but the variables that exist will always make it difficult to anticipate the outcome with confidence. Usually it will be more prudent to provide a clean soil layer, in which plant responses can be predicted.

11.4   ASSESSMENT OF RISKS TO PLANT POPULATIONS

Main Concerns

That plants in direct contact with a limited range of contaminants (List 11.1), or exposed to soil water containing high concentrations of these substances, will suffer phytotoxicity or concentrate substances harmful to human health.

These concerns will be of direct interest to individual home owners.

For these concerns, the worst-case assumption, that sensitive targets will always exist, has to be made.

[A]   Potential for future liability

Tick appropriate box

1. Does a fill layer mantle the site and is likely to extend to 2 m depths?
   (a) *Yes. Fill covers >50% of site.*   ☐
   (b) *Yes. Fill covers <50% of site.*   ☐
   (c) *No. Site appears to lack surface fills.*   ☐

2. If a fill layer exists, to the above depths, is it likely to contain List 11.1 substances in high concentrations?
   (a) *Yes. Past land uses indicate List 11.1 substances will occur at high concentrations.*   ☐
   (b) *No. List 11.1 substances not likely to be present.*   ☐

3. Where List 11.1 substances are likely to occur, are these probably plant-available? (Note pH control, origin of substances, etc.)
   (a) *Yes. Specify evidence below.*   ☐
   (b) *No. Specify evidence below.*   ☐

Q3. Evidence for plant availability/non-availability.

. . . .
. . . .
. . . .

4. If List 11.1 substances exist and could be more mobile, does a near-surface (<3 m) groundwater, or other transport mechanism, exist?

   (a) *Yes. Groundwater likely to be near surface and/or site loading could force fluids to rise.* ☐

   (b) *No. Groundwater not likely to be near surface, nor would loading be a likely transport mechanism.* ☐

5. If no fill layer exists, is it likely that spillages of List 11.1 fluids (e.g. wood preserving fluids) could have occurred on the site to a significant extent?

   (a) *Yes. Past land uses indicate that such fluids did exist on site.* ☐

   (b) *No. No such evidence exists.* ☐

6. Is there evidence that the site materials already have adversely affected plant populations?

   (a) *Yes. Vegetation abnormally absent/ stunted/reduced in diversity. Specify evidence below.* ☐

   (b) *No. No signs of any abnormality of vegetation density/type.* ☐

Q6. Evidence for adverse effects on plant populations.

. . . .
. . . .
. . . .

● ● ●

If the answers to all the above questions are "No", and if an initial review of available Site Investigation data confirms this, terminate plant population liability assessment at this point.

For other results, summarise the overall evaluation below. Justify clearly any decision to terminate plant population liability assessment.

. . . .
. . . .
. . . .
. . . .

LIST 11.1.

Contaminants which may *pose risks to plant populations*

Group A
Acidity or alkalinity   if outside of the pH range of 3 to 9

Group B
Arsenic	if > 40 mg/kg
Boron	if > 3 mg/kg
Cadmium	if > 15 mg/kg
Chromium (total)	if > 1000 mg/kg
Copper	if > 200 mg/kg
Mercury	if > 20 mg/kg
Nickel	if > 100 mg/kg
Zinc	if > 500 mg/kg
Cyanides (total)	if > 250 mg/kg
Phenols	if > 20 mg/kg
Sulphates	if > 2000 mg/kg
Tars (as PAHs)	if > 1000 mg/kg

Note: the effect of reducing pH is generally to make metallic contaminants more phytotoxic.
Individual variation within even a single cultivar is such that precise reliance cannot be placed on particular contaminant concentration values.

Group C
Landfill gases at root zones {	if methane > 1.5% by volume
	if carbon dioxide > 5% by volume
Carbon monoxide at root zones	if > about 1% by volume
Sulphur dioxide	if at any measurable concentrations

Scoring Options	SCORE
1. If either Group A and Group B contaminants occur at excessive concentrations (and particularly if complementary contaminants—such as the series copper, nickel and zinc—all occur), *and* if it can be taken that the contaminants are plant-available.	40
2. If Group C contaminants occur at excessive concentrations.	40
3. If *either* Groups A and B contaminants occur at high concentrations but in less plant-available forms *or* if Group C contaminants occur but just at the critical concentration levels.	20
4. If *either* Groups A and B contaminants occur but at relatively low concentrations which will not be largely plant-available *or* if Group C contaminants occur but below the critical concentrations.	10
5. If the contaminants can be shown to be non-plant-available.	0

[B] Probable magnitude of a future liability

1. Do List 11.1 substances exist above the appropriate ICRCL threshold trigger values? (Tables 1.3a and b)

   (a) *Yes. Specify details below.* 10

   (b) *To an extent. Specify details below* 5

   (c) *No. These substances absent or below the threshold trigger values.* 0

Q1. Presence of List 11.1 substances.

. . . .
. . . .
. . . .

---

2. If these substances do exist above the trigger values, are they likely to be plant-available? (note origins, pH control and other evidence such as leachability and see end of List 11.1 for scoring options)

   (a) *Yes. Specify evidence below.* 40–20

   (b) *To an extent. Specify evidence below.* 20–10

   (c) *No. Evidence (below) is that substances are not plant-available.* 10–0

Q2. Evidence of plant-availability.

. . . .
. . . .
. . . .

---

3. Does a local transport mechanism (e.g. near-surface groundwater) exist to move more mobile/soluble List 11.1 substances up to plant rooting zones? (*Note* also loading effects if List 11.1 substances are in mobile states)

   (a) *Yes. Specify details below. (e.g. groundwater < 3 m depths).* 20

   (b) *To an extent. Specify details below (e.g. groundwater < 5 m depths).* 10

   (c) *No. Plant rooting zones will not be affected by deeper contaminants. Specify details below.* 0

Q3. Local transport mechanisms.

. . . .
. . . .
. . . .

4. Are the local conditions such that   *(a) Yes. Materials at likely rooting*
   plant roots would tend to migrate   *depths have poor water-retaining*
   down into deeper contamination?   *characteristics and are unlikely to*
                                                 satisfy other plant requirements.
                                                 Specify below.   **20**

                                           (b) To an extent. Specify below.   **10**

                                           (c) No. Materials at likely rooting
                                                 depths have good water-retaining
                                                 characteristics and should satisfy
                                                 other plant requirements. Specify
                                                 details below.   **0**

Q4. Suitability of local conditions for plants.

. . . .

. . . .

. . . .

---

5. Is the Site Investigation adequate   *(a) Yes.*   **0**
   to permit the above questions to be
   answered with confidence?   *(b) To an extent. Specify deficiencies*
                                                 below.   **5**

                                           (c) No. Specify deficiencies below.   **10**

Q5. Information deficiencies.

. . . .

. . . .

. . . .

● ● ●

Advise Environment Division Director if risk evaluation has been biased
because of data inadequacies.

Overall Score   %   _____

[C] **Risk reduction achieved by site remediation**

Score (as positive values)

1. Has the remediation properly focused on preventing/reducing risks of phytotoxicity and plant uptake of contaminants?

   (a) Yes. — 10

   (b) Not entirely. Specify data below. — 5

   (c) No. Specify data below. — 0

Note: If the probable risks to plants have not been properly identified, then remediation benefits may not be entirely adequate.

Q1. Remediation focus adequacy.

....

....

....

Score only *one* of the remediation solutions—Q2 to Q4.

2.(a) Have all the List 11.1 substances been removed from the site?

   (a) Yes. Provide details below. — 70

   (b) Not entirely. Specify data below. — 35

   (c) No. Specify data below. — 0

Q2(a).

....

....

....

(b) Does quality control and/or post-remediation proof of this exist?

   (a) Yes. Specify evidence below. — 20

   (b) Not conclusively. Specify evidence below. — 10

   (c) No. Specify deficiencies below. — 0

Q2(b). Quality control/monitoring proof.

....

....

....

Score (as positive values

3.(a) If List 11.1 substances still exist on/in the site and below a 2 m thick clean surface layer, is this likely to prove an effective and permanent solution? (assume no landfill gas hazard persists. If this assumption incorrect reduce score given)

(a) Yes. Clean soil layer of fertile and good water-retaining type. No landfill gas hazard persists.   50

(b) To an extent. Specify details below.   25

(c) No. Clean soil type either poorly chosen for required fertility/water retention **or** could be recontaminated from near-surface polluted groundwater or exposed to continuing gas hazards. Specify below.   0

Q3(a).   Adequacy of clean surface layer.

. . . .

. . . .

. . . .

(b) If such a clean surface has been included for plant populations, does proof of its effectiveness exist? (e.g. quality control and/or monitoring data)

(a) Yes. Specify details below.   20

(b) Not conclusive. Specify details below.   10

(c) No. Specify details below.   0

Q3(b).   Quality control/monitoring proof.

. . . .

. . . .

. . . .

*Score (as
positive
values)*

4.(a) If List 11.1 substances still exist on the site and are covered only with a thinner soil layer, will this be an effective and permanent solution?

    (a)   *Yes, if garden areas are small (<50 m²) and if at least a 300 mm thick cover of water-retaining and fertile clean soil has been provided. This will only be satisfactory if no recontamination from near-surface polluted groundwater is possible.*   30

    (b)   *Position uncertain. Soil capping is > 150 mm but < 300 mm and/or is not of good water-retaining and fertile type. No concern over recontamination from near-surface polluted groundwater exists.*   15

    (c)   *No. Soil capping only 150 mm or less in thickness, and/or of permeable sandy nature, and/or could be recontaminated by near-surface polluted groundwater.*   5

    (d)   *No. As (c) but with additional continued risk of soil layer being affected by gases.*   0

(b)   If such a thinner soil capping has been installed, does proof exist of its effectiveness?

    (a)   *Yes. Quality control/post-reclamation monitoring data confirms the soil capping's quality. Details below.*   20

    (b)   *Not conclusive. Specify details below.*   10

    (c)   *No. Specify deficiencies below.*   0

Q4(b).   Quality control/monitoring data.

. . . .

. . . .

. . . .

• • •

Overall Score   %   _____

Remnant Risk   %   _____

(Section [B] score – Section [C] score)

11.5   ASSESSMENTS OF RISKS TO PLANT LIFE

The three-part assessment process detailed in Section 11.4 follows the usual pattern. The potential for future liability is defined by the presence of contaminant-rich surface fills, particularly where plant-availability is likely to be high and where a near-surface and contaminated groundwater occurs.

Experience suggests that—for other than very thin, sandy and acidic soils— that the ICRCL threshold concentrations (1987) for domestic gardens are far too cautious, and that harm to plants only occurs at or above the distinctly higher concentrations given in List 11.1.

The threshold values listed by the ICRCL for parks and playing fields are thus generally adopted, as those at which risks are probable. This choice of trigger values is defensible only on the basis that it agrees well with observed fact.

11.5.1   Site examples

All three example sites (Appendix I) certainly had (pre-reclamation) the potential to adversely affect plant populations.

The tipped wastes mantling Site No. 1 would have been expected to display medium concentrations of metallic and sulphate contamination (from the ash content) and periodically would have given rise to landfill gas concentrations likely to harm some plant life.

Similarly, the foundry fills on Site No. 2 (whilst not especially highly contaminated, except for the locally higher sulphate and mercury contents) would present poor rooting conditions for plants. The former oil terminal (Site No. 3) had locally excessively alkaline conditions and phenol and tarry contamination also was obvious, in addition to high levels of organic vapours.

In all cases, however, remediation should have ensured that concerns over risks to plant life were removed.

This achievement ought to have been most certain on Site No. 1, where the developer intended to remove all the contaminated materials. However, the failure to note the significance of probable spillages of timber preservative fluids, *and* to confirm that the exposed glacial clays were in fact uncontaminated, did lead to quite large-scale plant death. This outcome reflects no more than a lack of reasonable caution, and was emphasised by the importation of a particularly poor soil layer. This sandy soil's poor water retention and nutrient properties certainly encouraged plant roots to migrate down to the underlying glacial clays. Whilst these were compacted well enough to prevent no more than a few millimetres of plant root intrusion, the timber preservatives had coated this upper layer of the clays. Thus plant deaths occurred.

Site No. 2 encountered no problems from the soil contaminants. Partly this was due to the importation of a clay and nutrient-rich clean soil layer, and partly because lower layers were extremely thoroughly compacted, to highways standards. However, the proven existence of carbon dioxide concentrations (and depleted oxygen levels), far in excess of those known to be toxic to woody plants (Kozlowski, 1991), presented a much greater potential risk. Fortunately, long-term monitoring of gas flows (using flux box installations) was able to demonstrate that no carbon dioxide intrusion into the near-surface soils ever occurred. Thus, despite what would have appeared to be especially risky conditions for plant life, the site finally proved to be entirely free of this specific concern.

The final example site (No. 3) was initially especially at risk, if the benzene and oil vapours had been able to migrate into surface soils. However, this reclamation was completed to very high standards, and achievement standards were confirmed by rigorous quality assurance procedures.

A particular feature of the reclamation of Site No. 3 was the need to encapsulate tar-saturated sands and rocks and the spent oxides, which had accumulated in areas of the old gas works. These materials proved neither to be economically treatable or acceptable to local licensed waste disposal facilities. As these wastes had been of concern when they occurred spread through areas of the site, concerns must obviously be greater when it is intended to concentrate them in an encapsulation.

For an environmentally secure encapsulation, a prime aim must be to intercept rainfall which otherwise would come into contact with the encapsulated wastes and possibly produce contaminated leachate. This is *not* best achieved by attempting to prevent the entry of infiltration by (say) placing a surface layer of very impermeable clays above the wastes. Whilst this would appear, at first sight, to be a rational approach, experience reveals that surprisingly high amounts of annual rainfall can and do penetrate through such clay caps (Knox, 1991). In fact all that a clay cap achieves is to slow the rate of moisture migration, and this, in the long term, is no advantage whatsoever.

It is far better to:

(a) Compact the wastes, to give as few surfaces available for contact with any infiltrating water as possible, and so reduce to a minimum the leachability that is feasible. In this case, this was achieved by laying the wastes on a well-rolled clay substratum, and then compacting each thin waste layer (200 mm thick) to highways standards. Since tarry materials were placed as the lower and upper encapsulated layers, and the distinctly more leachable gas works wastes were placed within this sandwich of tarry and easily compacted fills, leachability was proved to be extremely low.

(b) Provide a layered capping (Figure 11.2) above the compacted wastes. At site No. 3 this capping had a surface layer into which rainfall could infiltrate, be stored and gradually migrate downwards. The lower capping layer, however, was chosen as a material (crushed concrete of very high saturated permeability) which would become water saturated as soon as moisture moved into it. This was then intercepted by outflow to the circumference drains, where it appeared (after a two-year lag) as clean groundwater.

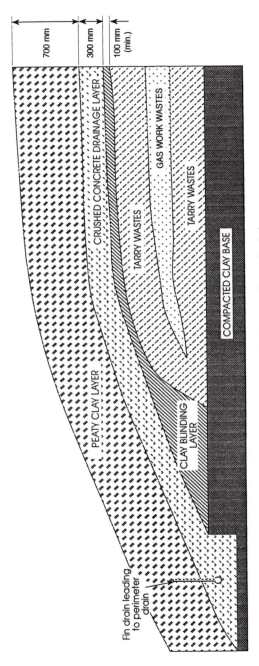

Figure 11.2. Encapsulation on oil terminal site

(c) Monitor the performance of the capping to ensure that it operates in accordance with design criteria and remains effective. Effectiveness is, of course, taken as a total interception of infiltrating rainfall and as no additional moisture reaching the encapsulated wastes.

Fuller details of the design approach are available in Cairney and Sharrock (1993).

Whilst the above procedure will adequately prevent the formation of leachates in the core of the encapsulation—and so remove concerns over groundwater or surface water pollution—it is necessary to consider how far plant populations could be affected.

The normal safeguard—provided by compacting the underlying peaty clay layer to a density well in excess of $1.4\,g/cm^3$—cannot easily be included, since this would compromise the design requirement that the uppermost layer of the encapsulation capping is a water storage medium into which infiltration can easily enter.

Thus it is necessary to install additional thicknesses of rooting soil and be careful to provide the type of soil which will be suitable for plant needs.

In the case of Site No. 3, this approach forced the provision of a 500 mm thick rooting soil layer (although the encapsulation was only to be a grassed public open space) *and* an additional 200 mm thickness for the underlying peaty clay cap (which otherwise need only have been 500 mm thick in total if the storage of two years' annual rainfall infiltration had been the sole design requirement). Additionally, care was taken to ensure that the quality of the surface soil layer was high. These precautions were deemed necessary to prevent plant root penetration which otherwise could have disrupted the designed multi-layered capping.

These three example sites indicate that avoiding contamination effects on plant populations is not especially difficult, if care is taken to physically prevent deeper movements of plant roots and also provide a suitable surface soil layer.

12

Risks to Human Health

12.1 CONCERNS OVER HEALTH IMPACTS

Exposure to a wide range of chemical compounds is strongly suspected to result in health damage. In consequence, developed nations have defined and enforce a range of allowable exposure limits for workforces in industrial and manufacturing situations (e.g. Health and Safety Executive, 1988).

The word "suspected" is necessarily employed above, since although a few contaminant exposures are certainly known to harm health (at least for more susceptible individuals), in most cases it is difficult or impossible to establish simple cause and effect relationships. Thus the conventional approach is to establish those exposure limits which produce no observed health effects. This, of course, implies that some meaningfully large factor of safety is included in each of the various allowable exposure limits—a point of practical importance to which further reference will be made (Section 12.3).

The concerns, in the various health and safety limits used by the developed industrial nations, focus primarily on the dangers of inhaling or ingesting chemical compounds, since this contact route is the one more likely to result in rapid health impacts.

12.2 WIDER ENVIRONMENTAL IMPACTS—THE DIFFICULTY IN ASCRIBING BLAME

If it is difficult to relate a particular level of exposure to chemical substances, to observable health reduction in the confined and monitored conditions of a modern manufacturing plant, the problem can only be greater if the exposure occurs in the wider environment around such plants. A recent and well-publicised case (Ghazi, 1993) makes the point.

There has been a deplorable worldwide flight of more polluting industries from the controls that exist in the developed nations. Relocation in Third World conditions is increasingly favoured as manufacturers seek freedom from restraints and also access to cheaper labour. Typical of such cases has been the influx of US corporations into the Matamoros area of the Mexico–USA border. Currently some 2100 such foreign companies are packed into one small and underdeveloped locality, and whilst about half produce large volumes of toxic wastes, little more than 300 appear to have obtained any permits to do so. Local regulators obviously find it difficult to show excessive zeal in enforcing worker safety and environmental protection ordnances (themselves far laxer than equivalent regulations in the USA or in Western Europe) when the economic benefits to the host nation are too important to risk.

The relocating companies gain real advantages (easy access to a First World market, yet a need only to accept Third World wage levels and rules) and the local community benefits directly from employment opportunities. But this is at the cost of air, water and soil pollution at levels which have not been seen in Western Europe for a century or more:

> factories openly dump huge mounds of calcium sulphate residues on street side walks; children play in pools of toxic green scum, and treacly black chemical wastes are tipped into ditches only yards from workers' shacks (Ghazi, 1993)

> . . . drainage water, containing the industrial solvent xylene at concentrations 6000 times the US drinking water standard, flows in a ditch outside of one car production factory (Ghazi, 1993)

The obvious scandal of this Dickensian industrial development forced the US and Mexican presidents to implement their 1992 Integrated Border Environmental Plan, and so it has to be taken that the relevant authorities did recognise that unacceptable health risks existed.

However, when an abrupt and totally localised rise in abnormal birth defects (including the fatal rare brain condition anencephaly) occurred in the early 1990s (and was also present in the neighbouring Texas town of Brownsville which lies downwind of the polluted Matamoros atmosphere), investigations found it impossible to identify any single causative factor. The Texas Department of Health and the US National Centre for Disease Control indeed suggested that diet and housing conditions could be contributing factors. Yet the incidence of anencephaly was four times that of the USA average, and similar in scale only to conditions in Matamoros itself!

If, in a case of this magnitude, it was not possible to ascribe blame, what chance is there with contaminated land, where concentrations are only a tiny fraction of those of toxic wastes? The answer, inevitably, has to be "almost none at all".

This view is supported by those few comparative studies on the health of communities living on contaminated land with similar populations from cleaner areas (e.g. Barltrop, 1986). Findings that:

> health inventories revealed few significant differences between the populations studied, and none which could be related to contamination (Barltrop, 1986)

make the point clearly.

Thus to claim that exposure to land contamination ever results in health deterioration goes well beyond what is currently provable. Yet public perceptions remain a factor of real importance, which no prudent developer should ignore. It may not yet be possible to ascribe any health problems to land contamination, but any suggestion that these might occur is likely to result in development blight and financial losses.

12.3 PROPOSED CONTAMINANT EXPOSURE GUIDELINES

If risks to human health from contaminated land are to be treated seriously, it is necessary to have guidelines which will distinguish safe from unsafe conditions. Additionally, these guidelines primarily should be expressed in terms of exposure to contaminant inhalation and ingestion, since these are the causative conditions of greatest probable concern. Other health risks (from gases and vapours and from eating crops that are contaminated) have been touched upon earlier (Chapters 9 and 11) and need not be considered in this context.

Most available health guidelines (e.g. Ontario Ministry of the Environment, 1986, and Table 12.1) choose to set standards on the total contaminant concentrations present in soils. But are these likely to be relevant for ingestion and inhalation risks? Surely the contaminant concentrations in dusts and vapours will be much more appropriate?

If this view is acceptable, it has to be recognised that a real difference exists between the presence of the same contaminant concentration in a loose, friable and fine-grained soil, and that in a densely compacted clay-rich matrix. In the first case, dust and vapour generation is likely, and so the opportunities for inhalation and ingestion of the contaminant cannot be dismissed; whilst in the second, contaminated dust can only occur if the soil is excavated, air-dried, broken up and then exposed to wind action.

Thus it seems more appropriate to utilise those allowable exposure limits (List 12.1) established for use in industrial premises (Health and Safety Executive, 1989). This approach has the logic of:

Table 12.1. Typical health-based contaminated land guidelines (after Ontario Ministry of the Environment, 1986)

Contaminant	Upper limits of normal soils (urban values in mg/kg)
Arsenic	20
Boron	15
Cadmium	4
Chromium	50
Cobalt	25
Copper	100
Lead	500
Mercury	0.5
Nickel	60
Zinc	500

- Utilising limits which have been derived from a much greater body of information than is available for any comparable contaminated land guidelines. Complete industrial populations have been monitored for decades and adverse health effects due to exposure to dusts and vapours are, at least, statistically well established.
- The limits do include factors of safety, and so extending the use of these values to populations other than those employed in industry is defensible. Industrial populations are likely to be adult, generally fit and so less susceptible than are very old, very young or sick people. However, with the included factors of safety, it seems reasonable to use the List 12.1 guidelines (8 hour exposure limits) for the population at large.
- Using limits which actually relate to the physical conditions (dusts and vapours) which in fact present ingestion and inhalation risks.

This approach, however, does call for a higher level of professional judgement than does using guidelines such as those cited in Table 12.1. Experience in the reclamation of a large number of contaminated sites, where air and dust monitoring took place, suggests that two conditions have to be fulfilled before the critical values of List 12.1 can occur:

- contaminant levels in the source soil have to be distinctly high, and certainly in excess of those established as safe thresholds for parks, playing fields and public open spaces (ICRCL, 1987), and
- the soils which surface a site have to be the types most easily disturbed and windblown. Experience indicates that finer sands and ashes (if heavily contaminated) are likely to be the most potentially hazardous materials.

The above emphasis on inhaling or ingesting contaminants as dusts may concern readers who believe that direct skin contact (with surface or near-surface contamination) could be a significant health hazard. Concerns, of course, often have been voiced that some young children have abnormal urges to eat non-food material, and that this (on a site where contamination is at site surface) could give rise to dangerously large chemical intakes.

These fears are believed to be overstated, particularly since a recent assessment, based on the information that no more than 10 such cases of abnormal appetites are reported each year in the UK (Becket, personal communication based on reported Department of Health information, 1994) calculates that the risk involved is trivial. Additionally, it is extremely improbable that any developer would leave (or indeed that any regulatory authority could permit) contaminated materials at or near the site surface. As a minimum (in the author's experience) a cover of clean soil is *invariably* provided. Thus the direct contact contaminant uptake route does appear to be of distinctly lower concern than is the risk from contaminated dusts, which of course are taken directly into internal human organs.

However, it is accepted that cases where only a thin clean cover above contaminated ground is provided do occur, and that deeper digging in such cases could bring contaminated materials back to the land surface. Should such cases occur, then the values cited in List 12.1 will not be appropriate, and assessment will necessarily have to be on the basis of the more usually employed total soil contaminant concentrations (List 12.2).

12.4 HUMAN HEALTH RISK ASSESSMENT

Main Concerns

That individuals might inhale or ingest a limited range of contaminants (List 12.1) or that direct contact with excavated contaminants will occur (List 12.2). Health risks created by gases/vapours and the eating of contaminated crops are addressed in Chapters 9 and 11 respectively.

These health concerns will be of direct interest to individual home owners.

For these concerns, the worst-case assumption, that sensitive targets will always exist, has to be made.

[A] Potential for future liability

Tick appropriate box

1. Does the site have a fill surfacing?
   - (a) *Yes. Fill covers >50% of site.* ☐
   - (b) *Yes. Fill covers <50% of site.* ☐
   - (c) *No. Fills absent.* ☐

2. Are any surface fills likely to contain List 12.1 or 12.2 contaminants at high concentrations?
   - (a) *Yes. Past land uses indicate that these substances will occur at high concentrations.* ☐
   - (b) *No. These substances not likely to be present.* ☐

3. Where List 12.1 or 12.2 substances are likely to exist in fills, are these directly accessible to people, or open to wind or water erosion?
   - (a) *Yes. Substances on site surface. Specify evidence below.* ☐
   - (b) *Yes. Site materials loose and likely to be wind or water erodible. Specify evidence below.* ☐
   - (c) *No. Specify evidence below.* ☐

Q3. Availability of health risk contaminants.

....

....

....

4. If no fill layer exists, is it likely that
   particular substances (e.g.
   asbestos) have been buried on the
   site?

(a) *Yes. Site adjacent to areas likely to
     have produced asbestos or similar
     wastes. Waste disposal authority
     information noted below.*   ☐

(b) *Position is not clear. Site adjacent
     to possible sources of asbestos, or
     similar wastes, but no actual
     evidence of their burial on this site.*   ☐

(c) *No. Likely sources of such
     contaminants do not exist locally.*   ☐

Q4. Waste disposal authority information.

....

....

....

---

5. If no fill layer exists, could the site
   ever have been affected by spillages
   of mobile List 12.1 or 12.2
   substances?

(a) *Yes. Tanks of such substances did
     exist on the site and so leakages
     could have occurred.*   ☐

(b) *No. No evidence to suggest such
     spillages.*   ☐

● ● ●

If the answers to all the above questions are "No" and if an initial review
of available Site Investigation data confirms this, terminate the human
health liability assessment at this point.

For other results, summarise overall evaluation below. Justify clearly
any decision to terminate human health liability assessment.

....

....

....

....

....

LIST 12.1

Contaminant concentrations and conditions which may pose inhalation/ingestion/contact risks to human health

Group	Contaminant		Conditions
A	Asbestos		If friable and windblown or if exposed to air. Critical level 0.2 fibres/ml of air
B (contaminant-rich dusts)	Arsenic		> 0.2 mg/m³ air
	Boron		>10.0 mg/m³ air
	Cadmium		> 0.05 mg/m³ air
	Chromium	if wind-	> 0.5 mg/m³ air
	Copper	blown—	> 1.0 mg/m³ air
	Cyanides	critical	> 5.0 mg/m³ air
	Lead	exposure	> 0.15 mg/m³ air
	Mercury	levels as	> 0.05 mg/m³ air
	Nickel	shown	> 1.0 mg/m³ air
	PCB		> 0.5 mg/m³ air
	Selenium		> 0.2 mg/m³ air
	Zinc		> 1.0 mg/m³ air
	Phenols	if ingested	>5 ppm
	Sulphates		>7.8 g ingested
	Tars		>trace occurrence

Note: Dust/inhalation/ingestion contact risks will *only* be significant if contaminants exist at well above ICRCL threshold values for parks, playing fields and open spaces, *and* if site surface (to relatively shallow depths) is contaminated. Concern will be greater if site surface materials are friable and easily eroded or dustblown.

Exposure levels noted above give relative importance of the listed contaminants.

Scoring Options	SCORE
1. If Group A contamination present in significant concentrations, at excavatable depths (i.e. in normal gardening or the repair of buried services), and in a friable condition.	40
2. If Group B contaminants are present (at excavatable depths) at concentrations well above ICRCL threshold values (parks, playing fields, open spaces) and in loose, friable soil matrices open to erosion and dust blow.	40
3. As (1) and/or (2), but with less easy migration possibilities. Note physical nature of site surface, existence of sealing layers, thicknesses of sealing layers, etc.	20
4. If Group A and/or Group B contaminants exist, but only just above ICRCL threshold values (parks, playing fields, open spaces). Migration will still be relatively easy due to loose/friable site surface.	20
5. If Group A and/or Group B contaminants exist only up to ICRCL threshold values and if migration possibilities are poor.	10
6. If contaminants are below the critical concentrations *and* erosion/dustblow prospects are very improbable.	0

LIST 12.2

*Contaminant concentrations which **may** pose contact risks if near-surface contaminated soil can be brought back to surface*

Arsenic	> 10 mg/kg
Boron	> 3 mg/kg
Cadmium	> 3 mg/kg
Chromium (hexavalent)	> 25 mg/kg
Chromium (total)	>600 mg/kg
Complex cyanides	>250 mg/kg
Free cyanides	> 25 mg/kg
Lead	>500 mg/kg
Mercury	> 1 mg/kg
PAHs	> 50 mg/kg
Phenols	> 5 mg/kg
Selenium	> 3 mg/kg
Thiocyanate	> 50 mg/kg

Scoring Options	SCORE
1. If the more toxic of the List 12.2 contaminants (arsenic, cadmium, mercury, polyaromatic hydrocarbons (PAHs), phenol and free cyanide) occur at high concentrations *and* at shallow depths (<0.5 m)	40
2. If these more toxic contaminants occur at high concentrations but at greater depths (≥1.0 m)	30
3. If only the less toxic of the List 12.2 contaminants occur at high concentrations and shallow depths (<0.5 m)	20
4. If only the less toxic contaminants occur but at greater depths (≥1.0 m)	10
5. If List 12.2 contaminants are at concentrations below those cited	0

[B] Probable magnitude of a future liability

Score (as negative values)

1. Do high concentrations of List 12.1 or 12.2 substances exist on the site or at shallow depths (<2 m) in it?

   (a) *Yes. Specify evidence below.*   10

   (b) *To an extent. Specify details below.*   5

   (c) *No. These contaminants do not occur, except at low concentrations. Specify details below.*   0

Q1. Presence of List 12.1 or 12.2 substances.

....
....
....

---

2. If these substances do exist, are they likely to be easily ingested/inhaled by people? (Note particularly if contaminants are in a fine-grained matrix and consider also the origin of the contaminants.) (see ends of Lists 12.1 and 12.2 for scoring options)

   (a) *Yes. Loose, friable and relatively fine-grained non-cohesive materials mantle the site and could give rise to contaminated dustblow, etc.*   40–20

   (b) *To an extent. Admixture with cohesive material reduces inhalation and ingestion risks.*   20–10

   (c) *No. Contaminants in a form which is likely to preclude inhalation and/or ingestion.*   10–0

3. If List 12.1 substances are capped with a cohesive cover, could the breaching of this give rise to ingestion/inhalation risks? Or could contact risks arise from deeper digging? (List 12.2)

   (a) *Yes. Specify details below.*   20

   (b) *To an extent. Specify details below.*   10

   (c) *No. Below cover, contaminants are not loose and/or are dispersed in cohesive soils.*   0

Q3. Ingestion/inhalation/contact risks.

....
....
....

4. If mobile types of the List 12.1 and   (a) *Yes. Mobile tars, etc. do exist*
   12.2 substances exist (e.g. tars,   *in the site and could move*
   organic fluids), can these pose   *upwards in response (for*
   health hazards? (relate to vapour   *example) to loading of the site.*
   risks; Section 9.2)   *Details below.*   **20**

                            (b) *Position uncertain. Details below.*   **10**

                            (c) *No. Mobile types of contaminants*
                                do not exist in the site.   **0**

Q4. Mobility risks.

. . . .

. . . .

. . . .

---

5. Is the Site Investigation data   (a) *Yes.*   **0**
   adequate to permit the above
   questions to be answered with   (b) *To an extent. Specify deficiencies*
   confidence?   *below.*   **5**

                            (c) *No. Specify deficiencies below.*   **10**

Q5. Information deficiencies.

. . . .

. . . .

. . . .

● ● ●

Advise Environment Division Director if risk evaluation has been biased
because of data inadequacies.

<u>Overall Score</u>   %   <u>   </u>

[C]   Risk reduction achieved by site remediation

Score (as positive values)

1. Has the remediation properly focused on preventing/reducing risks to human health?

    (a) Yes. — 10

    (b) Not entirely. Specify data below. — 5

    (c) No. Specify data below. — 0

Note: If the probable risks to human health have not been properly identified, then remediation benefits may not be entirely adequate.

Q1. Remediation focus adequacy.

. . . .

. . . .

. . . .

Score only *one* of the remediation solutions—Q2 to Q6.

2.(a) Have *all* the substances which could pose human health risks been removed from the site?

    (a) Yes. Details below. — 70

    (b) Not entirely. Specify data below. — 35

    (c) No. Specify data below. — 0

Q2(a).

. . . .

. . . .

. . . .

(b) If such removal has taken place, do quality control and/or post-remediation monitoring data exist to prove this?

    (a) Yes. Specify evidence below. — 20

    (b) Not conclusively. Specify evidence below. — 10

    (c) No. Specify deficiencies below. — 0

Q2(b).   Quality control/monitoring proof.

. . . .

. . . .

. . . .

3.(a) If the substances which could pose human health hazards still exist on the site, is their encapsulation and/or immobilisation likely to be effective and permanent?

(a) *Yes. Encapsulation and/or immobilised layer appears satisfactory and should not be recontaminated from below. (Have this situation checked by clean cover methodology and/or US EPA SITE process evaluation)* **50**

(b) *To an extent. Specify details below.* **25**

(c) *No. Encapsulation and/or immobilisation seem inadequate for long-term safety. Specify details below.* **0**

Q3(a).   Adequacy of encapsulation and/or immobilisation.

. . . .

. . . .

. . . .

(b)   If such encapsulation and/or immobilisation has taken place, does proof of its effectiveness exist?

(a) *Yes. Specify details below.* **20**

(b) *Not conclusive. Specify details below.* **10**

(c) *No. Specify details below.* **0**

Q3(b).   Proof of effectiveness.

. . . .

. . . .

. . . .

4.(a) Has the protection against human health risks consisted of only a soil capping (not a designed encapsulation)?

(a) Yes, and soil capping is thick enough (>1 m and preferably >1.5 m) to prevent exposure of contaminants in ordinary gardening. Local conditions would not recontaminate the soil cover.   40

(b) Yes, but capping is thinner (750 mm to 1 m) and/or could be recontaminated by local conditions (groundwater and/or earth pressures moving contaminants upwards).   20

(c) Yes, but capping is so thin (<500 mm) that casual digging could expose contaminants which are likely to pose health hazards.   10

(d) Yes, but capping is extremely thin (150 mm or less) and the contaminants below it would be likely to pose real health risks.   0

(b) If such a soil capping has been employed as the prime protection method, does evidence exist of its long-term effectiveness and durability?

(a) Yes. Specify details below.   20

(b) Not conclusive. Specify details below.   10

(c) No. Specify details below.   0

Q4(b). Effectiveness/durability of soil capping.

....
....
....

5.(a) If the main protection method has been to cap the site with hard surfacing, does this seem likely to be effective? (no areas of open soil left on site) (applies only to sites being utilised for non-domestic and commercial purposes)

(a) Yes. Total capping with concrete/tarmac laid to a proper specification. Site under control of a single management agency and buried services are in clean reservation areas.   40

(b) Yes. Site is properly capped, but no single management agency exists to ensure that uncontrolled excavations do not occur.   30

(c) No. Capping not convincingly specified/does not cover entire area/site not under single management agency's control.   20

*Score (as
positive values)*

(b)   If the site is hard surfaced, does   *(a)*   *Yes. Specify details below.*   20
      evidence exist of the quality of the
      surfacing?   *(b)*   *Not conclusive. Specify details
                                                   below.*   10

                                           (c)   *No. Evidence is not available.
                                                   Specify deficiencies below.*   0

Q5(b).   Quality control/monitoring proof.

. . . .

. . . .

. . . .

6.(a) If the site has been reclaimed by   *(a)*   *Yes. Specify details below.*   50
      a particular mix of the above
      techniques, does this seem likely   *(b)*   *Not entirely. Specify details
      to be effective and durable?   and reasons below.*   25

                                           (c)   *No. Specify details and reasons
                                                   below.*   0

Q6(a). Mix of remediation and likely effectiveness.

. . . .

. . . .

. . . .

(b)   In any such case does quality   *(a)*   *Yes. Specify details below.*   20
      control and/or post-remediation
      proof of effectiveness exist?   *(b)*   *Evidence not conclusive.
                                                   Specify details below.*   10

                                           (c)   *No. Specify details below.*   0

Q6(b). Quality control/monitoring proof.

. . . .

. . . .

. . . .

                                    Overall Score   %   _____

                                    Remnant Risk   %   _____

                           (Section [B] score – Section [C] score)

12.5 RISK ASSESSMENTS

The two prerequisites—high contaminant concentrations at site surface and loose and easily blown soils—indicate that obvious ingestion or inhalation risks will be very rare on reclaimed sites. It simply is not credible that any competent developer would produce such an unsatisfactory result, or that control bodies would permit this.

Thus risks—if they are to occur—will be limited to situations where deeper contamination still exists. This might be intersected (by later maintenance or gardening work) or be brought up by rising soil water in long hot droughts, and could then generate direct contact risks in addition to those from contaminated dusts.

The three example sites (Appendix I) should, after reclamation, have been free of concerns that human health risks could occur. However, information deficiencies do leave remnant concerns, particularly for one area on Site No. 1.

The light industrial site (Site No. 1) overall never had high enough contamination concentrations to suggest that any health risks were possible, and the removal of the site's ash surfacing should have dispelled any concern which might have been voiced. There is, of course, the area where timber preservatives were spilled, but this (whilst of concern when risks to plant populations were assessed) is not significant in terms of dust inhalation or ingestion. The spilled fluids now exist as mineral coatings on the surface and parting planes in the stiff glacial clays, and these materials are most unlikely ever to generate dusty atmospheres.

However, direct contact risks could arise, if a home occupier in this area chose to excavate deeply for (say) a garden pond or a home extension and brought contaminated clay to the surface. This might well expose materials likely to be both acidic and rich in bio-available metals. Thus additional information certainly is required on the contaminant condition of the upper few millimetres of the glacial clays.

The former iron foundry (Site No. 2) initially had more worrying conditions. Highish cadmium, lead, phenol and sulphate concentrations in friable ashy fills could have given rise to contaminated dusts. However, the removal of the upper fills, their replacement by a very well-compacted 2.0 m-thick clean cover, and the further provision of 300 mm of clayey soils in garden areas removes almost all possible concerns. Even if home owners do excavate deeply in gardens, contaminated dusts or direct contaminant contact should never occur. The sole doubt—that of the clean cover gradually becoming contaminated by moisture rising from the site's groundwater—exists only because no effort was made to identify whether the groundwater was in fact contaminated. This information deficiency has been noted in other risk assessments and indicates the widespread consequences that can arise from a deficiency in site investigation.

The final site (the former oil terminal) quite obviously posed very real health hazards in its unreclaimed state. This was especially the case above the old gas works, and around the demolished oil tanks. However, in this development, reclamation care had been especially thorough. Contaminated fluids were pumped out, soils were excavated and separated into clean and contaminated types, and post-remediation checks revealed that only demonstrably clean soils now cover the sandstone rock. Thus, for the areas to be used for housing development, there can be no concern that impacts on residential health could arise.

However, one salient aspect of this reclamation was the on-site encapsulation (Figure 11.2) of wastes (oil and tar-soaked soils and cyanide-rich materials) which could neither be treated or taken off-site. Concentrating these wastes in one spot obviously implies the potential for future risk. Designing such an encapsulation so that it is environmentally secure is not especially difficult (Cairney, 1993b), but it has to be noted that such designs only preclude unwanted movements of moisture and any contaminants dissolved in these fluids.

Other encapsulation failure mechanisms, however, can occur—i.e.:

- because of slope instability,
- due to erosion by surface run-off, and
- if ground settlement occurs.

These, either singly or together, could be especially worrying, since the outcome could be the exposure of broken-up waste layers to wind and rain, and so the occurrence of contaminated dusts and vapours.

It thus will be necessary *not only* to check that the encapsulation design has fully addressed these potential failure events and that adequate safety factors have been included, *but* also to undertake routine annual inspections of the encapsulation so that any necessary corrective maintenance is initiated before any ingestion or inhalation health hazards can arise.

12.6   CONCLUSIONS

It is still difficult or impossible to demonstrate direct impacts on human health even when uncontrolled exposure to very high contaminant concentrations has taken place. Only in a relatively few cases (e.g. with asbestos fibres) is a cause and effect relationship well established.

When exposure is to the relatively low contaminant concentrations which typify most contaminated soils, it is unlikely that any demonstrable health reductions will ever be proved. However, public perception can be more severe than scientific assessment, and developments on formerly contaminated sites could be blighted if health concerns were voiced.

Avoiding such concerns is not difficult if attention is directed to the causative conditions (contaminant-rich dusts) of real importance and if reclamation quality assurance is demonstrated. Attention must also be paid to cases where deeper contamination still exists and could be intersected by predictable gardening or excavation.

13

Conclusions

It is indisputable that contaminated land is a legacy which ought to be rectified as quickly as possible. Equally it is obvious that other public health, welfare and environmental improvements are far more important and pressing. Social deprivation is still much too widespread, even in richer nations, for the cleaning up of most old industrial sites ever to be given more priority than (say) improving educational opportunities, removing slum housing, or ensuring a wider access to modern medical care. Contaminated land can in some circumstances be a *potential* problem, but other social problems are much more certain to create unacceptable consequences.

Thus, leaving the enormous clean-up costs to the state purse is a practical nonsense. Even one small nation, such as the Netherlands, has located more than 120 000 suspect sites, each of which calls for investigation costs and possibly remedial fundings. The Dutch, uniquely, had taken the view that national resources would have to be the prime source of land remediation investments, but now have found that this burden is far too onerous, and are seeking other solutions.

It can, of course, be argued that making the "polluter pay" is the answer. The state may have to find the initial monies, but (on this argument) will recover those once guilty parties are identified. But is this really a sensible proposition? It sounds very proper, criminals certainly ought to discharge their obligations; though how they are to be found when they often ceased trading or went bankrupt decades ago is seldom made clear. Any suggestion of pursuing their descendants seems much more likely to enrich lawyers than to compensate the public purse for funds that have already been expended.

In conditions—so far mainly restricted to the UK—where the private-sector has been permitted to reclaim contaminated sites and restore these for productive re-use, this dilemma has been avoided. Public funds have generally not been required and reclamation standards have been ensurable by a range of legislative and administrative controls. Since the private-sector (if the controls do not erode too far the final profits when reclaimed land is resold) is capable of particularly rapid reclamations, land remediation in the UK has achieved rates not seen elsewhere.

The only problem with this approach is that potential land sales profits have to be significant, and this translates into allowing more sensitive re-uses for reclaimed sites. In today's conditions domestic housing and light industrial developments guarantee the best final profits, and these re-uses increase human exposure to any risks which contaminated land remediation has failed to remove.

This essentially is the basic justification for promoting a consistent risk assessment methodology, and for arguing that seven distinct risk categories each should be evaluated.

Land scarcities are a factor of life in Western Europe. Pressures on greenbelts have become so intense that it can be difficult to spot where one town ends and its neighbour begins. A sprawl of housing estates continues to absorb more and more open land. Yet old industrial centres are often left derelict and sparsely populated. Surely this cannot be a reasonable policy.

The reclamation of most old industrial sites is neither especially difficult or particularly offputting to private developers. Certainly there are problem sites, which ought to be tackled by the state for the welfare of the population and the wider environment, but these are few in number. For most contaminated areas, the essential safeguard is to ensure that acceptable safety and reclamation quality is achieved, and to do this does call for risk assessments which are both affordable and consistent.

Appendix I

Examples of Contaminated Sites, all Reclaimed for Housing Re-Use

SITE NO. 1 EXCAVATION AND REMOVAL SOLUTION
LIGHT INDUSTRIAL SITE—CENTRAL SCOTLAND

Site location and layout

The site lies on the outer fringes of Glasgow. Figure AI.1 shows the site layout.

Site history

Till 1920s—agricultural use.

1920s to 1945—timber storage yard occupied the entire site.

1945—a northern section of the site was utilised for the tipping of demolition rubble into a shallow (3 m deep) gravel pit.

1945 to 1963—timber treatment processes were established on south-east area of the site.

1963 to 1978—northern half of site utilised as a car breaking yard. Southern half was left unused.

1978 to 1986—uncontrolled waste tipping took place over the entire site.

Site topography
(pre-reclamation)

The land was essentially level and some 1.0 m higher than surrounding land. Low heaps of wastes occurred on the northern frontage with the main road.

No surface watercourses exist. The closest surface water body (200 m to the north) is entirely culverted below the housing estate.

Natural geology
(see Figure AI.2)

Glacial clays are recorded (up to 8 m thick) and localised lenses of gravel are known to occur within these clays. Below are Coal Measures Shales (unmined) to depths of at least 25 m.

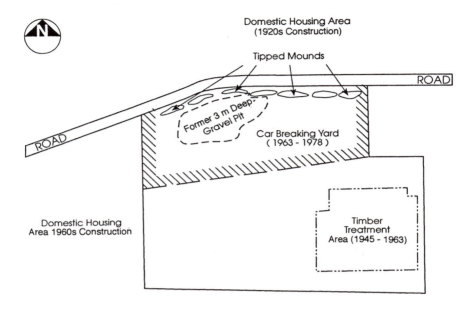

Figure AI.1. Site No. 1: former light industrial site (0.79 ha)

Site investigation
(January 1989)

16 trial pits, at 25 m spacings, and to depths of 3.0 m.

Two boreholes, to 10.5 m depths, for foundation exploration.

Chemical investigation was limited to the testing of a near-surface sample of the tipped materials, exposed in each trial pit (Table AI.1).

No chemical investigations were carried out on samples of the underlying glacial clays.

No groundwater analyses were carried out.

No landfill gas or vapour investigations were carried out.

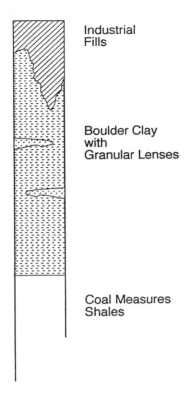

Industrial
Fills

Boulder Clay
with
Granular Lenses

Coal Measures
Shales

Figure AI.2. Vertical section through Site No. 1

Table AI.1. Proven chemical contamination of the surface fills (mg/kg except for pH values)

pH	6.06 to 7.65
Sulphides (as H_2S)	<1
Sulphates (2 : 1 water : soil extract)	253 to 2036
Toluene-extractable content	1170 to 21 120
Phenols	1.1 to 17.4
Cyanide (free)	<2
Arsenic (total)	8.1 to 34.3
Cadmium (total)	<1 to 7.6
Chromium (total)	2.1 to 49.3
Copper (plant-available)	3.6 to 478
Lead (total)	12 to 653.2
Nickel (plant-available)	0.8 to 36
Zinc (plant-available)	2.8 to 623
Ammonium	9.2 to 39.7
Asbestos	ND

ND = not detected.

Site investigation had revealed the presence of a pervasive and thin (~ 1.0 m thick except where mounds of tipped wastes occurred on the northern site frontage) capping of combustion ashes and domestic wastes.

The former gravel pit was located and found to be filled with well-compacted and apparently clean demolition rubble (brick and stone debris), which appeared to have resulted from the post-war urban repair programme.

Perched groundwater, as minor volumes, did occur in gravel lenses in the boulder clays. The underlying Coal Measures Shales proved to be dry to depths of 10.5 m.

Reclamation choice

The developer decided to remove all contamination concerns, by having all fills excavated and removed to an off-site tip. This choice removed the usual requirement for sub-surface gas investigations.

The glacial clay surface and the top of the rubble-filled gravel pit were exposed, and house units were founded directly into the exposed surface.

A sandy soil was imported to cap proposed garden areas (to 300 mm thickness).

No formal quality controls were exercised, and reclamation proceeded without specialist supervision or inspection.

Planned re-use	Semi-detached housing and gardens.
Later problems	Extensive plant die-back occurred in several gardens in the south-eastern corner of the site. This forced later re-investigation (Table AI.2), which also located spilled oils and vapours in the infilled gravel pit.
Comment	A small and apparently quite simple site, of the type very attractive to small and medium-sized building firms, and located in an area of housing demand.

Table AI.2. Additional contamination investigations (June 1992)

Location	Depth (m)	Results
Plot 23 (south-east corner of site)	0.100	Plant-available copper 74 mg/kg
(all samples had acidities of 4.5	0.200	Plant-available copper 136 mg/kg
or less)	0.400	Plant-available copper 274 mg/kg (total chromium concentrations also increased markedly with depth)
North of site over infilled gravel pit	0.500	Oxygen 12%, carbon dioxide 3%
	1.000	Oxygen 2%, carbon dioxide 11%
	1.500	Oxygen <1%, carbon dioxide 15%

SITE NO. 2   CLEAN COVER SOLUTION
FORMER IRON FOUNDRY—SOUTH-WEST MIDLANDS

Site location and layout	South-western fringe of Birmingham and adjacent to greenbelt. Site layout shown on Figure AI.3.
Site history	Till 1860s—agricultural use.
	1860s to 1800—unknown industrial use.
	1880s to 1960s—iron foundry.

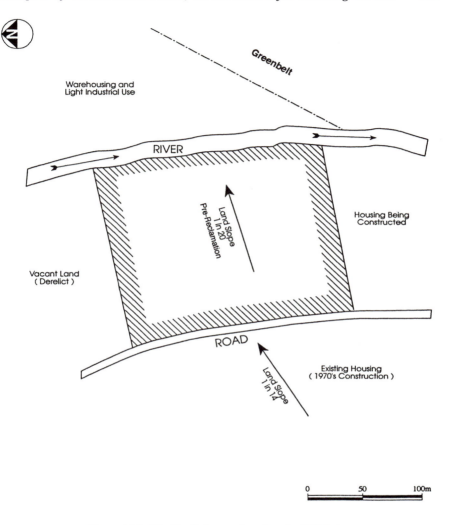

Figure AI.3. Site No. 2: former foundry site (3.21 ha)

1964—foundry demolished.

1964 to 1986—site derelict and capped with concrete (roadways and foundation slabs).

Site topography
(pre-reclamation)

Surface slopes to adjacent river. Land to south slopes steeply towards the site.

Local river relatively clean, and its quality is improving as improvements occur to various sewage treatment works.

Natural geology
(see Figure AI.4)

Water-bearing sands and gravels (up to 4 m thick) are the uppermost natural deposits. Below these are unmined Coal Measures Shales (42 m thick) above Carboniferous Limestone rocks. The limestone was mined in the 19th century.

Site investigation
(October 1987)

43 trial pits, at 30 m spacing, and to depths of 4.2 m.

Eight boreholes to prove the depth and condition of the Coal Measures rocks.

Iron foundry fills which occurred as a 4.0 m thick layer were sampled at three depths in each trial pit and chemically analysed (Table AI.3).

Groundwater occurred in the fills at 3.0 m depth, and also in the underlying sands and gravels. No sampling or analysis of groundwaters was carried out.

Landfill gas investigations were limited to a gas spike test survey, to depths of 1.0 m into the exposed foundry fills (Table AI.4).

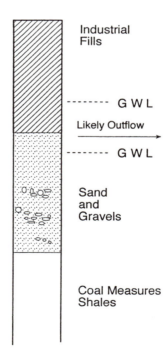

Figure AI.4. Vertical section through Site No. 2

Table AI.3. Proven chemical contamination of the foundry fills (mg/kg except for pH values and loss on ignition)

pH	7.06 to 8.04
Arsenic (total)	0.1 to 10.2
Cadmium (total)	0.5 to 29.7
Chromium (total)	8.8 to 33.1
Lead (total)	51.7 to 4355
Mercury (total)	<0.1 to 10.4
Copper (plant-available)	8.1 to 49.3
Nickel (plant-available)	0.7 to 10.73
Zinc (plant-available)	51.0 to 179
Sulphates (acid soluble)	659 to 4375
Sulphides	<1.0 to 6.4
Phenols	0.3 to 8.4
Cyanide (free)	0.06 to 24.7
Toluene-extractable content	633.0 to 1148
Loss on ignition	13 to 44.2%
Free oils	<1000
Asbestos	ND

Note: higher concentrations typify upper layers of foundry fills.
ND = not detected.

Site investigation revealed that iron foundry wastes invariably existed over the site. Chemical investigations indicated that the upper 1.5 m of these fills were distinctly more contaminated than were the lower layers. The upper layers also contained wood and fabric scraps, which gave rise to low concentrations of methane (up to 0.2% by volume) and of carbon dioxide (up to 4.2% by volume).

Reclamation choice

It had been hoped that all the contaminated fills could have been removed, but the quantities present proved to be too great. Thus it was decided to remove only the more contaminated materials (to 2.0 m depths) and then lay a clean cover over the remaining fills. A source of clean quarry sand and gravel provided materials for the cover.

During reclamation, the site was terraced to reduce the previous gradient to the river and a final land surface 1.5 m above mean river water level was constructed.

A clay-rich soil (300 mm thick) was provided in garden areas.

Considerable care was taken to avoid any contaminated fluids entering the river during reclamation, to confirm the uncontaminated nature of the clean cover materials. To prove that compaction of the cover had been satisfactory, a quality assurance system was operated.

Table AI.4. Gas concentrations in the site

	CH_4 (%)	CO_2 (%)	O_2 (%)
1. Initial surveys (spike tests to 1.0 m depths)	<0.1	<3.6	>18.0
2. Later surveys (boreholes to 15 m depths)	nil	up to 15%	less than 1%

House units were founded on slabs laid on the compacted clean cover.

Planned re-use   Domestic housing, some gardens and some blocks of flats.

Later concerns   A potential purchaser of the site voiced concerns over the continued existence of foundry fills at 2.0 m depths in the site. A minor re-investigation did prove that reclamation work had been as detailed above, but also indicated the presence of high carbon dioxide concentrations and very low oxygen levels in areas of the site (Table AI.4).

Comment   A large and quite complex site where the main reclamation aims have been to avoid any polluted run-off entering the adjacent river and to providing a provably safe clean capping. No note was taken of water pollution via the near-surface groundwaters. This was a normal attitude in the late 1980s, when wider environmental degradation was not the concern which it is today.

SITE NO. 3   CONTAMINANT REMOVAL SOLUTION
OIL TERMINAL SITE—NORTH-WEST ENGLAND

Site location and layout   Adjacent to coast. Bounded by a clean stream and by land still in agricultural use. Site layout shown on Figure AI.5.

Site history   Till 1880—agricultural use.

1880 to 1914—small gas works in eastern central area of the site.

1914 to 1950s—Royal Navy fuel depot.

1950s to 1981—private-sector oil storage depot.

1981—site demolished and levelled.

Site topography (pre-reclamation)   Flat and some 2.0 m higher than the surrounding farmland. Concrete cut-off wall separates the site from the stream to the west.

Natural geology (see Figure AI.6)   Fluvio-glacial sands (up to 3 m thick and water-bearing) overlie massive sandstones.

Site investigation (summer 1990)   3 m deep trial pits on a 20 m grid.

Ten deeper (20 m) boreholes to prove groundwater conditions in the sandstone rock.

Ground radar geophysical survey to map out contact of sands and the underlying sandstone, and identify low spots on the rock surface.

Groundwater occurred in the sands and showed floating oils plus oil emulsions.

Free oil product proved to be thickest (10 to 15 mm thick) where low spots in the sandstone surface existed.

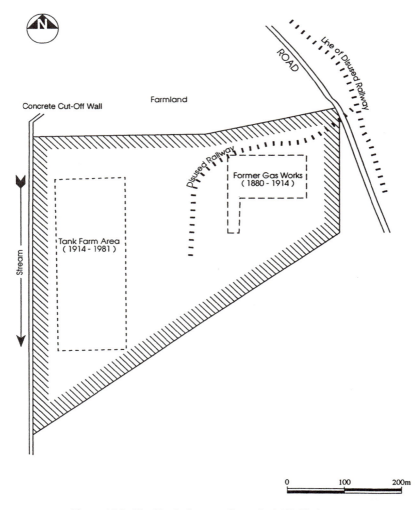

Figure AI.5. Site No. 3: former oil terminal (18.8 ha)

Oil contamination restricted to tank farm area and to the former loading area, adjacent to the rail head.

Sands heavily oil contaminated in these two areas. Rock surface (especially at low spots) was heavily coated with denser oil faction. Oils had penetrated only some 20 mm into the sandstone rock.

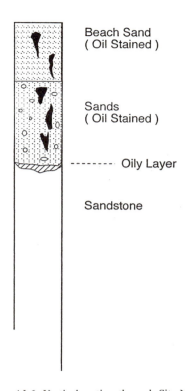

Beach Sand
(Oil Stained)

Sands
(Oil Stained)

- - - - - - - -   Oily Layer

Sandstone

Figure AI.6. Vertical section through Site No. 3

Gas works wastes and contaminated groundwaters in fluvio-glacial sands adjacent to the former gas works.

Upper 2.0 m of site composed of imported beach sands. Presumably laid in 1914, since they entirely cover the old gas works.

Chemical analyses of gas works wastes (Table AI.5).

Chemical analyses of oil-soaked sands (Table AI.6).

Chemical analyses of benzene air pollution (Table AI.7).

Site zoning revealed three areas of concern, namely: the former gas works, the tank farm area, and the fuel loading area at the rail head. The remainder of the site proved to be essentially clean.

Groundwater/oil plume proved to move in a south-west direction from the former tank farm area.

Reclamation choice

The presence of liquid, gaseous and solid contamination forced a reclamation choice designed to minimise contaminant migration and so reduce benzene levels that the site could be excavated without risks of air pollution.

Table AI.5. Chemical composition of gas works wastes

(a) *Solid materials* (all as mg/kg except pH and calorific value)
pH	5.7 to 8.4
Total cyanide	<2.4 to 6400
Free cyanide (as CH)	<2.5 to 340
Toluene-extractable content	500 to 18 000
Coal tar derivatives (PAHs)	<100 to 12 000
Mineral oil	<500 to 4000
Total phenols	<2.5 to 8.2
Acid soluble sulphates	600 to 16 000
Sulphides (as H_2S)	<1.4 to 64
Total arsenic	8.6 to 18
Total lead	50 to 430
Total zinc	70 to 410
Calorific value	<5000 kJ/kg

(b) *Waters* (all mg/litre except pH)
pH	3.4 to 8.3
Total cyanide	<0.05 to 10.0
Free cyanides	<0.05 to 0.32
Toluene-extractable content	7 to 33 000
Phenols	<0.05 to 7.1
Sulphates	350 to 5000
Sulphides	<0.02 to 4.4
Total lead	<0.02 to 0.35
Total zinc	<0.02 to 1.9

Table AI.6. Chemical composition of oil-contaminated areas

(a) *Solid materials* (all as mg/kg except pHs)
pH	6.7 to 9.1
Total sulphates (acid soluble)	80 to 1260
Total phenols	2 to 186
Toluene-extractable content	4000 to 34 600
Total PAHs	28 to 2200

(b) *Liquid samples* (all as mg/litre except pHs)
pH	6.1 to 9.2
Total sulphates	340 to 1517
Total phenols	<0.1 to 23.7
Toluene-extractable content	46 to 99 500
Total PAHs	<0.1 to 6670

Table AI.7. Benzene air pollution when 3 m-deep excavations were carried out

Time (minutes)	Excavation depth (m)	Benzene concentration in air (mg/litre)
0	0	0
12	3	137 to 102
15	3	64 to 41
20	3	20 to 12
30	3	10 to >5

Note: Benzene concentrations in air vary with wind condition and are lower in higher winds.

The reclamation sequence was:

(a) Isolation of areas known to be contaminated by oil spillages and where floating free oils existed. Sheet pile cut-offs utilised.
(b) Removal of contaminated groundwaters and free oils from these isolated areas. Treatment of discharged water (filtration/de-emulsification/final biological treatment). Pumping continued until benzene concentrations in monitoring boreholes had declined to ~5 mg/litre of air.
(c) Excavation of contaminated soil and the upper 50 mm of contaminated rock from within the isolated areas. Washing of excavated soil to reduce oil content. Less treatable materials stockpiled for future encapsulation. Quality control procedures in force.
(d) Excavation of old gas works site with stockpiling of contaminated soil and gas works wastes for future on-site encapsulation.
(e) Regrading of site and final sampling of site materials and vapour surveys to prove reclamation effectiveness.
(f) Encapsulation of wastes (from stages (c) and (d)) on north-eastern corner of the site.
(g) Provision of 500 mm of fertile and water-retaining soils in areas where vegetation cover was required.

Planned re-use   Housing, gardens and public open spaces.

Comment   A large and difficult site, where significant health risk and environmental hazards existed. Reclamation management and quality control were particularly good.

Appendix II

Risk Assessments of the Example Sites

(A) DESK STUDY EVALUATIONS (PRE-RECLAMATION POTENTIAL FOR FUTURE LIABILITIES)

Site No. 1
Risks to surface water	Nil
Risks to groundwater	Low to Nil
Risks of area-wide air pollution	Low
Risks from gases/vapours	Medium to Low
Risks to construction materials	Low
Risks to plant populations	Medium to Low
Risks to human health	Low

Site No. 2
Risks to surface water	Medium
Risks to groundwater	Low to Nil
Risks of area-wide air pollution	Low
Risks from gases/vapours	Low
Risks to construction materials	Medium to Low
Risks to plant populations	Medium to Low
Risks to human health	Low

Site No. 3
Risks to surface water	High
Risks to groundwater	High
Risks of area-wide air pollution	Medium to Low
Risks from gases/vapours	High to Medium
Risks to construction materials	High to Medium
Risks to plant populations	High to Medium
Risks to human health	Medium

(B) PROBABLE MAGNITUDE OF FUTURE LIABILITIES (PRE-RECLAMATION)

Site No. 1
Risks to surface water	0
Risks to groundwater	0
Risks of area-wide air pollution	− 15
Risks from gases/vapours	− 35
Risks to construction materials	− 25

| | Risks to plant populations | -40 |
| | Risks to human health | -15 |

Site No. 2	Risks to surface water	-20 to -55*
	Risks to groundwater	0
	Risks of area-wide air pollution	-15
	Risks from gases/vapours	$-10/-25$†
	Risks to construction materials	-40
	Risks to plant populations	-45*
	Risks to human health	-25

Site No. 3	Risks to surface water	-70
	Risks to groundwater	-75
	Risks of area-wide pollution	-40
	Risks from gases/vapours	-40
	Risks to construction materials	-50
	Risks to plant populations	-70
	Risks to human health	-50

*Evaluation influenced by lack of information on local groundwater.
†First score refers to methane risk, second to that from carbon dioxide.

(C) RISK REDUCTION ACHIEVED (POST-RECLAMATION)

Site No. 1	Risks to surface water	N/N
	Risks to groundwater	N/N
	Risks of area-wide air pollution	$+100$
	Risks from gases/vapours	$+60$
	Risks to construction materials	Still unclear*
	Risks to plant populations	$+50$†
	Risks to human health	$+85$

Site No. 2	Risks to surface water	Still unclear‡
	Risks to groundwater	N/N
	Risks of area-wide air pollution	$+100$
	Risks from gases/vapours	$+70$
	Risks to construction materials	$+45$‡
	Risks to plant populations	$+60$‡
	Risks to human health	$+50$‡

Site No. 3	Risks to surface water	$+100$
	Risks to groundwater	$+100$
	Risks of area-wide air pollution	$+100$
	Risks from gases/vapours	$+100$
	Risks to construction materials	$+100$
	Risks to plant populations	$+100$
	Risks to human health	$+100$

N/N = no need to consider
*Assessment influenced by lack of information on the contaminant state of the glacial clays in the wood treatment area.
†Assessment after corrective action had taken place in the areas of plant death.
‡Assessment influenced by lack of information on local groundwater.

(D)   REMNANT LIABILITIES (POST-RECLAMATION)

Site No. 1	Risks to surface water	Nil
	Risks to groundwater	Nil
	Risks of area-wide air pollution	Nil
	Risks from gases/vapours	Probably Nil
	Risks to construction materials	Still unclear*
	Risks to plant populations	Probably Nil
	Risks to human health	Probably Nil*
Site No. 2	Risks to surface water	Still unclear
	Risks to groundwater	Nil
	Risks of area-wide air pollution	Nil
	Risks from gases/vapours	
	Risks to construction materials	
	Risks to plant populations	probably Nil
	Risks to human health	
Site No. 3	Risks to surface water	Nil
	Risks to groundwater	Nil
	Risks of area-wide air pollution	Nil
	Risks from gases/vapours	Nil
	Risks to construction materials	Nil
	Risks to plant populations	Nil
	Risks to human health	Nil

*Doubt must still persist until the chemical condition of the upper surface of the glacial clay is proved.

References

Alloway, B. J. and Morgan, H. (1986) The behaviour and availability of Cd, Ni, and Pb in polluted soils. In Assink, J. W. and Van Den Brink, W. J. (eds) *Contaminated Soils*. Martinus Nijhoff, Dordrecht, pp. 101–114.

American Concrete Institute (1991) Proposed Revision of the Guide to Durable Concrete, Report by ACI Committee 201. *ACI Materials Journal*, September/October 1991, pp. 544–582.

Angus Environmental Ltd (1991) *Review and Recommendations for Canadian Interim Environmental Quality Criteria for Contaminated Sites*. Canadian Council of Ministers of the Environment, Ottawa.

Anon. (1989) Cleaning up sites with on-site process plants. *Environmental Science and Technology*, **23**, 912–916.

Anon. (1991a) *Environmental Protection (Prescribed Processes and Substances) Regulations*. HMSO, London.

Anon. (1991b) *Methane—Facing the Problems*. Symposium, Nottingham University, UK.

Anon. (1993a) Crisis developing over environmental measurements. *Environment Business*, 30 June edition. Information for Industry Ltd, London, p. 1.

Anon. (1993b) Legal aid sets up pollution and clean-up test cases. *Environment Business*, 28 July edition. Information for Industry Ltd, London, p. 1.

Anon. (1993c) Boys to sue Rechem in pollution test case. *Environment Business*, 6 October edition. Information for Industry Ltd, London, p. 3.

Anon. (1993d) Has DoE been moving Part B goalposts? *Environment Business*, 20 October edition. Information for Industry Ltd, London, p. 1.

Anon. (1993e) Air pollution treatment. *Environment Business Technology Guide*, October. Information for Industry Ltd, London, pp. 1–20.

Anon. (1994a). Licence defence succeeds. *Environment Business*, 23 February edition. Information for Industry Ltd, London, p. 6.

Anon. (1994b) Causing polluting matter to enter controlled water. *The Times*, 4 August edition.

Barltrop, B. (1986) Evaluation of cadmium exposure from contaminated soil. In Assink, J. W. and Van Den Brink, W. J. (eds) *Contaminated Soils*. Martinus Nijhoff, Dordrecht, pp. 169–181.

Barry, D. (1991) Hazards in land recycling. In Fleming, G. (ed.) *Recycling Derelict Land*. Thomas Telford, London, pp. 28–63.

Barry, D. L. (1983) *Material Durability in Aggressive Ground*. Construction Industry Research and Information Association, Report No. 98, CIRIA, London.

Beauchamp, G. S. (1993) Establishing new landscapes. In Cairney, T. (ed.) *Contaminated Land: Problems and Solutions*. Blackie Academic & Professional, London, pp. 191–211.

Becket, M. Personal communication. Estimation of contaminant ingestion risks to "pica" children in the UK.

Bell, C. E. *et al.* (1989) State of research and regulatory approach of state agencies for clean-up of petroleum contaminated soils. In Calabrese, E. J. and Kostecki, P. T. (eds) *Petroleum Contaminated Soils*—Vol. 2.

Bridges, E. M., Chase, D. S. and Wainwright, S. J. (1979) Soil and plant investigations since 1967. In Bromley, R. D. F. and Humphrys, G. (eds) *Dealing with Dereliction*. University College of Swansea, pp. 23–42.

British Gas (1992) *Environmental Review*. British Gas plc, London, UK.

British Standards Institution (1971) *Building Drainage*. Code of Practice CP 301: 1971, Milton Keynes.

British Standards Institution (1988) *Draft for Development—Code of Practice for the Identification of Potentially Contaminated Land and its Investigation*. DD 175: 1988, BSI, London.

Building Control Regulations (1991) *Approved Document C2*. HMSO, London.

Building Research Establishment (1991) *Sulphate and Acid Resistance of Concrete in the Ground*. Digest No. 363, BRE, Garston, UK.

Cairney, T. (1993a) Environmentally secure cappings for landfills. In Proceedings Conference *Discharge Your Obligations*, Kenilworth, Warwickshire, November 1993, pp. 123–130.

Cairney, T. (1993b) Clean cover technology. In Cairney, T. (ed.) *Contaminated Land: Problems and Solutions*, Blackie Academic & Professional, Glasgow, pp. 84–110.

Card, G. B. (1993) *Protecting Developments from Methane*. Construction Industry Research and Information Association, Report C.P/8, London.

Childs, K. A. (1985) Treatment of contaminated groundwater. In Smith, M. A. (ed.) *Contaminated Land*. Plenum Press, New York.

CIRIA (1995) Decommissioning and demolition. Vol. II in *Remedial Treatment for Reclaiming Contaminated Land*. Construction Industry Research and Information Association, London.

Clark, K. (1993) The AERIS model—its use in risk assessment of contaminated land. In *Site Investigation for Contaminated Land*. IBC Technical Services, Ltd, London.

Crowhurst, D. (1987) *Measurement of Gas Emissions from Contaminated Land*. Building Research Establishment, Fire Research Station, Boreham Wood, Herts, UK.

de Bruijn, P. (1992) Biotreatment in soil remediation. In *Contaminated Land—Policy, Regulation and Technology*. IBC Technical Services, London.

Department of Environment (1985a) *Landfill Sites: Development Control*. Circular 17/87, HMSO, London.

Department of the Environment (1985b) *Building Regulations, Part C*. HMSO, London.

Department of the Environment (1986) *Landfilling Wastes*. Waste Management Paper No. 26, HMSO, London.

Department of Transport (1991) *Annual Abstract of Transport Statistics*, HMSO, London.

Dobson, M. C. and Moffat, A. J. (1993) *The Potential for Woodland Establishment on Landfill Sites*. Department of the Environment, HMSO, London.

European Communities (1975) *Directive on the Quality Required of Surface Water Intended for the Abstraction of Drinking Water*. 75/440/EEC. Official Journal of the European Communities. L 194/26.EC.

Evelyn J. (1661) *Fumi Fugium: Or the Inconvenience of the Air and Smoke of London*.

Failey, R. A. and Bell, R. M. (1991) Behaviour of pollutants in soils. In Davies, M. C. R. (ed.) *Land Reclamation—An End to Dereliction?* Elsevier Applied Science, London, pp. 215–227.

Ferguson, C. and Denner, J. (1993) Soil remediation guidelines in the UK—a new risk based approach. In Proceedings of *Developing Clean-up Standards for Contaminated Soil, Sediment and Groundwater*, Washington DC, 10–13 January 1993, pp. 205–211.

Flavin, R. J. and Harris, R. C. (1991) Contaminated land: implications for water pollution. *Journal Institution of Water and Environmental Management*, **5**(2), 529–533.

Ghazi, P. (1993) America's deadly border. *Observer Magazine*, 12 December edition, pp. 16–21. Guardian Newspapers Ltd, London.

Griffin, R. P. (1988) *Principles of Hazardous Materials Management*. Lewis Publishers, USA.

Harris, M. R. (1987) Recognition of the problem. In Cairney, T. (ed.) *Reclaiming Contaminated Land*. Blackie & Son Ltd, Glasgow, pp. 1–29.

Haynes, M. H. and Zubiate, P. C. (1973) *Compressive Strength of 67 Year Old Concrete Submerged in Sea Water*. Technical Note N-1308, Naval Civil Engineers Laboratory, Point Hueneme, California, USA.

Health and Safety Executive (1988) *Approved Code of Practice: Control of Substances Hazardous to Health*. COSHH Regulations 1988, HMSO, London.

Health and Safety Executive (1989) *Occupational Exposure Limits*. Guidance Note EH 40/89, HMSO, London.

Heilman, P. (1981) Root penetration of Douglas fir seedlings into compacted soil. *Forestry Science*, **27**, 660–666.

Henriksen, C. F. and Stoltzner, E. (1993) Chloride corrosion in Danish road bridge columns. *Concrete International*, August 1993, 55–60.

John, M. K. and Van Learhoven, C. J. (1976) Differential effects of cadmium on lettuce varieties. *Environmental Pollution*, **10**, 163–173.

Knox, K. (1991) Water management at landfills: water balance and practical aspects. National Association of Waste Disposal Contractors course notes (unpublished), Coventry.

Kozlowski, T. T. (1991) Soil aeration, compacting and flooding. In Kozlowski, T. T., Kramer, P. J. and Pallardy, S. G. (eds) *The Physiological Ecology of Woody Plants*. Academic Press, San Diego.

La Goy, P. K., Nisbet, I. C. T. and Schulz, C. O. (1989) Endangerment assessment for the Smuggler Mountain Site—Pitkin County, Colorado. In Paustenback, D. J. (ed.) *Risk Assessment of Environmental Hazards*, John Wiley & Sons, Chichester.

Leone, I. A., Flower, F. B., Gilman, E. F. and Arthur, J. J. (1979) *Adapting Woody Species and Planting Techniques to Landfill Gas Conditions—Field and Laboratory Investigation*. US EPA Report 600/2-79-128.

Lepp, N. W. and Harris, M. R. (1980) A strategy for evaluation of soil covering techniques to reduce trace metal uptake by soft fruits and vegetables. In Proceedings Conference *Reclamation of Contaminated Land*, Eastbourne. SCI, London, pp. C7/1–C7/9.

HM Inspectorate of Pollution (1989) *The Control of Landfill Gas Waste*. Waste Management Paper No. 27, HMSO, London.

Hoomans, J. P. and Stellingwerf, J. W. (1982) *Operation Lekkerkerk West*. Pt. Civiele Techniek Vol. 1, pp. 3–44.

ICRCL (1983) *Guidance on the Assessment and Redevelopment of Contaminated Land*. Guidance Note 59/83, 1st Edition, Interdepartmental Committee on the Redevelopment of Contaminated Land, Department of the Environment, London.

ICRCL (1987) Guidance on the Assessment and Redevelopment of Contaminated Land. Guidance Note 59/83, 2nd edition, Interdepartmental Committee on the Redevelopment of Contaminated Land, Department of the Environment, London.

Institute of Wastes Management (1990) *Monitoring of Landfill Gas*. IWM, Northampton.

Institution of Civil Engineers (1991) *Inadequate Site Investigation*. Thomas Telford, London.

International Standards Organisation (1993) *Soil Quality Sampling—Pt 5: Sampling Strategies for the Investigation of Soil Contamination of Urban and Industrial Sites*. ISO CD10381 (3rd draft).

Lewin, K. and Bradshaw, K. (1993) *Pollution Potential of Contaminated Sites: A Review.* Research and Development Note 181, National Rivers Authority, Bristol, UK.

Lord, A. (1991) Options available for problem solving. In Fleming, G. (ed.) *Recycling Derelict Land.* Thomas Telford Ltd, London, pp. 145–195.

Marshall, T. C., Dubinsky, M. and Boutwell, S. (1989) A risk assessment of a former pesticide production facility. In Paustenback, D. J. (ed.) *Risk Assessment of Environmental Hazards.* John Wiley & Sons, Chichester.

Matta, Z. G. (1992) Chloride and corrosion in the Arabian Gulf environment. *Concrete International*, **14**(5), 47–48.

Ministry of Housing, Spatial Planning and the Environment (1994) *Environmental Quality Objectives in the Netherlands.* Risk Assessment and Environmental Quality Division, Ministry of Housing, Spatial Planning and the Environment, The Hague, Netherlands

Moen, J. E. T. *et al.* (1986) Soil protection and remedial actions: criteria for decision making and standardisation of requirements. In: Assink, J. W. and Van den Brink, W. J. (eds) *Contaminated Soils.* Martinus Nijhoff, Dordrecht, pp. 441–448.

Murdock, L. J. and Brook, K. M. (1979) *Concrete Materials and Practice*, 5th edition. Edward Arnold, UK.

National Rivers Authority (1992) *Policy and Practice for the Protection of Groundwater.* National Rivers Authority, Bristol.

Nederlands Normalisatie-Institut (1990) *Soil—Investigation Strategy for Exploratory Surveys.* Guidance Note MNV 5740, NNI, Delft.

Ontario Ministry of the Environment (1986) *Clean-Up Criteria for Soils.* Ontario, Canada.

Orchard, D. F. (1979) *Concrete Technology—Properties of Materials* (Vol. 1), 4th edition. Applied Science Publishers, London.

Parry, G. D. R. and Bell, R. M. (1987a) Types of contaminated land. In Cairney, T. (ed.) *Reclaiming Contaminated Land.* Blackie & Son, Glasgow, pp. 30–38.

Parry, G. D. R. and Bell, R. M. (1987b) Landscaping and vegetating reclaimed sites. In Cairney, T. (ed.) *Reclaiming Contaminated Land.* Blackie & Son, Glasgow, pp. 213–222.

Peterson, P. J., Benson, L. M. and Zieve, R. (1981) Metalloids. In Lepp, N. W. (ed.) *Effect of Heavy Metal Pollution on Plants.* Applied Science Publishers, London, pp. 279–342.

Raloff, J. (1980) Disaster on 99th Street. *New Scientist*, 19 June issue, pp. 298–300.

Rees, J. F. and Grainger, J. M. (1982) Rubbish dump or fermenter? *Process Biochemistry*, Nov./Dec. 1982, pp. 41–44.

Robinson, N. and Grayson, R. (1990) Natural methane seepages in the Lancashire Coal Field. *Land and Mineral Surveyor*, July 1990.

Rodricks, J. V. (1992) *Calculated Risks—Toxicity and Human Health Risks of Chemicals in our Environment.* Cambridge University Press.

Rorsch, A. (1986) Foreword. In Assink, J. W. and Van Den Brink, W. J. (eds) *Contaminated Soils.* Martinus Nijhoff, Dordrecht.

Sleeman, W. (1990) Environmental effects of the utilisation of coal mining wastes. In Rainbow, A. K. M. (ed.) *Reclamation, Treatment and Utilisation of Coal Mining Wastes.* A. A. Balkema, Rotterdam, pp. 65–76.

Smith, M. A. (1991) Data analysis and interpretation. In Fleming, G. (ed.) *Recycling Derelict Land.* Thomas Telford, London, pp. 88–144.

Smith, M. A. (1995) *Site Investigation and Assessment—Vol. III Remedial Treatment for Contaminated Land.* Construction Industry Research and Information Association Research Project 443. CIRIA, London.

Smith, M. V. (1993) Landfill gases. In Cairney, T. (ed.) *Contaminated Land: Problems and Solutions.* Blackie Academic & Professional, Glasgow, pp. 160–190.

Thornton, I. (1986) Metal contamination of soils in UK urban gardens: implications for health. In Assink, J. W. and Van Den Brink, W. J. (eds) *Contaminated Soils.* Martinus Nijhoff, Dordrecht, pp. 203–211.

216 *References*

Traves, L. (1992) Applying risk assessment concepts to evaluate alternative uses of contaminated industrial properties. Proceedings of *Decommissioning, Decontamination and Demolition Conference*, IBC Technical Services, London.

Tretheway, K. R. and Chambers, J. (1988) *Corrosion*, 5th edition. Longman Scientific and Technical, UK.

Trevelyn, G. M. (1964) *Illustrated English Social History—Vol. IV*. Pelican Books Ltd, Middlesex, UK.

Twardowska, I. and Szczepanska, J. (1990) Transformations of chemical compositions of pore solutions in coal mining wastes. In Rainbow, A. K. M. (ed.) *Reclamation, Treatment and Utilisation of Coal Mining Wastes*, A. A. Balkema, Rotterdam, pp. 177–186.

Twardowska, I., Szcezepanski, A. and Tejeszerski, J. (1990) Prognosis of contaminants leaching from colliery spoils. In Rainbow, A. K. M. (eds) *Reclamation, Treatment and Utilisation of Coal Mining Wastes*, A. A. Balkema, Rotterdam, pp. 153–164.

Ward, R. S., Williams, G. M. and Hill, C. C. (1993) Changes in landfill gas composition during migration. In Proceedings Conference. *"Discharge Your Obligations"*, Kenilworth, Warwickshire, pp. 381–392.

Williams, G. M. and Hitchman, S. P. (1989) The generation and migration of gases in the subsurface. In Proceedings Symposium *Methane—Facing the Problems*, Nottingham University, UK.

Index

DATE DUE

NOV. 1 2 1996			
DEC 0 1996			
DEC 11 '9?			
MAY 0 4 1999			
NOV 0 9 1999			
GAYLORD			PRINTED IN U.S.A.